Bird Migration

Bird Migration

Chris Mead

Facts On File Publications
460 Park Avenue South
New York, N.Y. 10016

Acknowledgements

Photographic Acknowledgements

Colour
Ardea, London: 195 bottom, 216, R. J. C. Blewitt 197,
Su Gooders 113, C. R. Knights 15 bottom, Robert T.
Smith 64 bottom, David & Katie Urry 156, Richard
Vaughan Endpapers; British Museum, London: 22 top;
Bruce Coleman, Uxbridge: 65 top, Jane Burton 27
bottom, R. Carr 30 top, S. B. Dawson 22 bottom,
Inigo Everson 114, Jeff Foott 14, Robert Gillmor 30
bottom, J. L. G. Grande 97, P. A. Hinchliffe 4–5,
Gordon Langsbury 195 top, Wayne Lankinen 18 top,
Leonard Lee Rue 18 bottom, M. F. Soper 10 bottom,
R. Tidman 10 top, Simon Trevor 65 bottom, Joseph
Van Warmer 11 top, 15 top, 26 top, Peter Ward 11
bottom, 27 top, Roger Wilmshurst 23, 67, Gunter
Ziesler 26 bottom; Chris Mead: 128 top, 194 top, 194
centre, 194 bottom.

Black and White
British Library, London: 29, 31; Hamlyn Group
Picture Library: 35, 186 bottom; Marconi Research
Centre, Chelmsford: 201; Hugh Miles, Wimborne: 186
top; Ralph Morse, New Jersey: 128 bottom, 138;
Natural History Survey Division, Illinois: 145 top left,
145 top right, 145 bottom; John Topham Picture
Library, Edenbridge: 50, 90, 91, 93 top, 101, R. J.
Tulloch 84–85, W. Puchalski 93 bottom; University of
Exeter: 25.

Maps and diagrams by The Hayward Art Group
Drawing on page 114 by John Busby

Published in the United States of America in 1983 by Facts
On File, Inc., 460 Park Avenue South, New York, N.Y. 10016.

Published in Great Britain by Country Life Books, an imprint
of Newnes Books and distributed by The Hamlyn Publishing
Group Limited.

Library of Congress Cataloging in Publication Data

Mead, Chris.
  Bird migration.
  Bibliography: p.
  Includes index.
  1. Birds–Migrations.   I. Title.
QL698.9.M43   1983   598.252'5   82-15385
ISBN 0-87196-694-8

Printed and bound by Graficromo s.a., Cordoba, Spain

References cited in diagrams

Where appropriate, the source material for line
drawings has been acknowledged. However, in some of
the composite drawings, derived from several widely
disparate sources, acknowledgement has not been
practical.

Aschoff, J. *Proceedings XIV Int. Orn. Congr.* (1967)
Baldaccini *et al J. Comparative Physiology*, **99** (1975)
Bergman & Donner *Acta. Zool Fenn.* (1964)
Bernis, A. J. *et al Ardeola* (1973)
Berthold, P. & Querner *Science* **212** (1980)
Bragg, A. M. *Living Bird* (1967)
Cochran, Montgomery & Graber *Living Bird* (1967)
Cochran, W. W. *Animal Orientation and Navigation*
  NASA SP 262 (1972)
Eckhardt, R. C. *Minimum time paths and the migration
  of the Arctic Tern* Office of Naval Research, Harvard
  (1969)
Emlen, S. T. *Auk*, **84** (1967)
Emlen, S. T. & Emlen, J. T. *Auk*, **83** (1966)
Gwinner, E. *J. Ornithology* **110** (1969)
  Proceedings XV Int. Orn. Congr. (1972)
Gwinner, E. & Wiltschko, W. *J. Comparative
  Physiology* **125** (1978)
Hale, W. G. *Distribution of the Redshank in its Winter
  Range Zool. J. Linn. Soc.* **53** (1973)
Keast, A. *Migrant Birds in the Neotropics* Smithsonian
  Press
Keeton, W. *British Birds* **72** (1979)
  Scientific American **23** (1968)
McDonald *Animal Orientation and Navigation* **NASA
  SP262** (1972)
Mayr, E. *History of North American Bird Fauna;
  Wilson Bull.* **58** (1946)
Mengel *Living Bird* **3** (1964)
Merkel, F. W. & Frome, H. G. *Naturwiss* **45** (1958)
Mewaldt, L. R. *Science* **146** (1964)
  Western Bird Bander **38** (1963)
Michener, M. C. & Walcott, C. *Journal of
  Experimental Biology* **47** (1967)
Moreau, R. E. *The Palearctic-African Bird Migration
  Systems* (1972) Academic Press
Perdeck, A. C. *Ardea* **55** (1958)
Pettigrew, J. D. from Schmidt-Koenig, K. & Keeton,
  W. T. (eds) *Animal Migration, Navigation and
  Homing*, Springer Verlag (1978)
Salomensen, F. *The Evolutionary Significance of Bird
  Migration Dan. Biol. Medd.* **22** (1955)
Schmidt-Koenig, K. *Cold Spring Harbour Symp.* **25**
  (1960)
Sharrock, J. T. R. *Rare Birds in Britain and Ireland*
  Poyser (1976)
  Scarce Migrant Birds in Britain and Ireland Poyser
  (1974)
Svensson, L. *Identification Guide to European Passerines*
  Swedish Museum of Nat. Hist. (1975)
Wagner *Rev. Suisse Zool.* **75** (1968, 1972)
Ward, P. *Ibis* **113** Academic Press
Williams *et al Animal Orientation and Navigation*
  NASA SP262 (1972)
Wiltschko, W. *J. Comparative Physiology* **109** (1976)
Wiltschko. W. *Z. Tierpsychol*, **25** (1968)
Wiltschko & Wiltschko *J. Ornithology* **117** (1976)
Yeagley, H. L. *J. Appl. Phys* **22** (1951)

Contents

Introduction

Migration may be defined as a regular movement between areas inhabited at different times of the year. Bird migration therefore encompasses a very wide range of species from all parts of the world, particularly the temperate and cooler zones where there are clear seasonal climatic differences. The usual understanding is that the bird which breeds north of the equator will make a southern movement to spend the winter in a warmer area – this is certainly a common strategy but by no means the only one. Most people also tend to think of a particular bird species as being a 'migrant' – like the familiar Swallow – and others as being sedentary. This very simple view is not supported by the birds for, within very many species, the populations in some areas may be sedentary whilst others, from a different part of the range, may be long-distance migrants. Indeed there are populations of many species where a part of the population migrates to warmer areas for the winter whilst others remain – the phenomenon of partial migration.

This book is concerned with all sorts of bird migration as they affect birds of the Holarctic, a region comprising North America and Eurasia. Most of the birds concerned breed and move within the area

(or further south) during the winter. A few seabirds are the only examples of birds breeding south of the equator that regularly migrate to spend their winter (the Holarctic's summer) north of the equator. The relative lack of temperate zone land-masses south of the equator is undoubtedly the main reason for this discrepancy.

It is clearly quite easy when thinking in terms of human movements to appreciate that the migration of a summer visitor to our temperate area is comparable to a person taking an extended holiday in a Mediterranean resort. This superficial analogy conceals the major ecological truth about migrant birds: they are as natural a part of the local bird community in the regions they visit as any of the sedentary birds in each of the areas. In fact there are many species which are only in their breeding areas for three or four months of the year and may be seen in their wintering area for almost twice as long. Some of the larger species, that do not breed until they are two or three years old, may even spend their first summer still in the 'wintering' area and not travel northwards at all. Far from considering them as northern species which happen to winter in the south it might be better to consider them as southern species

6

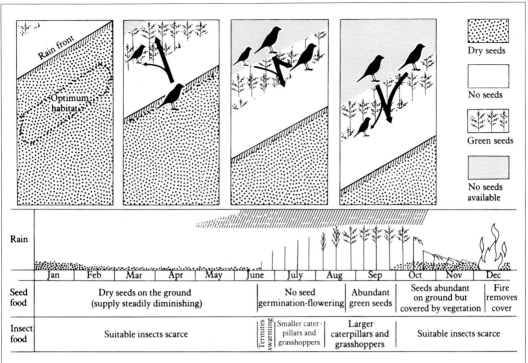

	Jan	Feb	Mar	Apr	May	June	July	Aug	Sep	Oct	Nov	Dec
Seed food	Dry seeds on the ground (supply steadily diminishing)					No seed germination-flowering		Abundant green seeds		Seeds abundant on ground but covered by vegetation		Fire removes cover
Insect food	Suitable insects scarce					Termites swarming	Smaller cater-pillars and grasshoppers	Larger caterpillars and grasshoppers		Suitable insects scarce		

Dry seeds

No seeds

Green seeds

No seeds available

After P. Ward

Seasonal movements of Quelea in Africa. Initially the birds are feeding on dry seeds (1) but, as the rains pass by and the grasses begin to seed, about six weeks later, early birds move to the new food source (2). Six weeks later the big breeding colonies are formed in optimal areas (3) but, six weeks later again (4) the rains have passed through the main Quelea area there are only small colonies on poorer feeding grounds.

which happen to breed in the north! The swifts provide a good example, since in North America the Chimney Swift and in Eurasia the Common Swift seldom reach their more northern breeding grounds until May and have nearly all begun their return journey by mid-August. The local birdwatchers in their breeding area consider them to be an essential part of their bird fauna but each Swift probably spends twice as long in its wintering area – over the Amazonian forest region of Brazil for the Chimney Swift or the farms and forest of East Africa for the Common Swift – than on their breeding grounds.

Let us stay with these swifts as representatives of long-distance migrants. It is easy to see that they are both very accomplished

flying machines and gather their food in flight. Although when searching for food it is possible for them to journey long distances in a relatively short time, it is a fact that the aerial plankton on which they depend is only in their breeding haunts during the short, warm summer. However, despite the short duration, this represents a good source of nourishment for any species of bird able to exploit it. The swifts are not the only birds able to hawk flying invertebrates; in both the Old and New World a variety of hirundines also exploit the resource. The swifts are regularly to be seen flying and foraging higher than the hirundines and are therefore able to exploit a different food supply. Thus, in the case of the swifts, the migrant is able to breed in an area where it could not possibly survive during the winter, not simply through the direct effect of climatic difference but because of the total lack of suitable food. Nonetheless the underlying constraint is the seasonality of the environment: throughout the study of migration this factor is the one which recurs.

In the case of the swifts the main aspect of seasonality is the temperature difference but, in other circumstances, the particular effect of seasonality which determines the movement of birds may be different. For example, within Africa and Australia bird movements are very often determined by the occurrence of rains. In Africa these tend to be predictable and in the form of bands of precipitation gradually travelling across the dry areas. These rains promote a tremendous growth of vegetation, breeding of invertebrates and, ultimately, setting of vegetable seed. Many bird species move with these fronts. In Australia the rains in the dry areas are not as predictable as in Africa. However, when rains do occur, vast and spectacular changes take place and birds of all sorts may appear and breed. Even such groups as ducks, waders and other water-birds arrive and breed on the temporary lakes and rivers that are formed. These changes to the environment may have a great effect on the migrants from temperate areas which visit Africa and Australia but the movements of endemic birds within those continents are outside the scope of this book. In any case the movements of the weavers in Africa and the adventitious breeding birds of the Australian Outback are nomadic rather than being truly migratory: they are not regular movements between alternate areas inhabited at different times of the year.

The Holarctic migrants are an immensely varied group of birds; they include the exhilaratingly agile swifts and the apparently weak-flying wrens. In fact they breed in every imaginable habitat and feed on every possible sort of food. The numbers of birds involved on every autumn migration is so huge as to be almost unbelievable. Reg Moreau, in his classic book, *The Palaearctic-African Bird Migration Systems* (1972), estimates that about five billion migrants leave Europe and Asia to winter south of the Sahara

in Africa each autumn. Within the Old World there are countless millions more long-distance north-south migrants making the return trip to India, south-east Asia and even Australia. In the New World many long-distance migrants from North America reach South America. However, these birds only make up a part of the full range of Holarctic migrants for there are also the arctic duck, geese and waders which are winter visitors to the temperate regions where the human population is concentrated. Add to these the east-west migrants (like the thrushes of Eurasia), the really short-distance migrants and the seabirds of the Pacific and Atlantic Oceans and the full variety and gigantic scale of bird movements begins to make itself felt.

That migration is such a common strategy amongst birds indicates that it must be a successful proposition. The tendency to migrate would quickly be lost from any population of birds if it led to unacceptably high losses. Indeed, if there were such high losses, unless part of the population was remaining on the breeding grounds the population would dwindle and eventually become extinct. In the most extreme cases the advantages of migration are clear – tiny insectivorous birds are able to do very well in raising their young in high latitudes where the short summer provides a tremendous flush of insect life. Even in more temperate areas it is obviously not possible for insectivorous species to survive as their food supply dwindles almost to nothing during the cold of the winter. There are basically two options open to a temperate region insectivore determined not to move away during the winter cold.

The first option is to specialize. Even when the snow is lying there are some insects and other invertebrates available within thick clumps of vegetation where there may be more warmth and where the snow has not penetrated. This is how the Carolina Wren of North America and the Dunnock of Europe are able to remain in apparently inhospitable places although, with both species, some may migrate. Another means of specializing, followed by the chickadees of America and their close relatives, the titmice of Europe, is to search for the dormant invertebrates within their winter hiding places. These busy, agile birds, continually searching the bark of trees and other likely places, have two over-riding priorities which force them into efficient food finding during the depths of winter. First the cold can only be combatted by consuming food which can be 'burnt' within their bodies to produce heat – actually finding a really snug roosting place can help a lot by reducing overnight heat loss. The very length of the long winter nights provides the second problem because, at exactly the time that they will need the maximum amount of food, because of the coldness and length of the night, the available daylight for feeding is at a minimum. The Brown Creeper (America) or Treecreeper (Europe) goes one better than the tits and chickadees in that it has a very fine, long beak adapted for reaching the dormant invertebrates in cracks and crevices. This strategy reaches perfection in the woodpeckers. Their formidable beaks enable them to uncover the large grubs of insects within their otherwise safe hiding places.

The second option is to change feeding habits. This has recently been demonstrated for an interesting small bird, the Bearded Reedling, of the southern Palearctic. The bird, superficially looking like a titmouse, is actually closely related to the tropical babblers. During the summer it lives in reed-beds and raises several broods of young feeding exclusively on insects and other invertebrates. Early in the autumn it undergoes its main moult, whilst continuing to feed insectivorously. However, even in the north of its range in Holland and England, many remain in the reed-beds throughout the winter. It was initially thought that it was able to remain an insectivore by finding dormant arthropods after the fashion of the true tits. Observations in the early nineteenth century that it was able to feed on the *Phragmites* seeds during the winter were discounted. However, it is now clear that this is indeed the case and the remarkable bird undergoes a change in its digestive system to enable it to cope with the very different problems of digestion of the hard seeds, as opposed to the soft-bodied animals of the summer diet. From late September the gizzard gets thicker and more muscular so that the seeds may be crushed. Until December this feature is not reversible but a warm spell in February or March can cause the summer gizzard to return – to the great detriment of the birds if a cold snap follows. This strategy is generally an efficient way of remaining on the breeding grounds in the winter but the bird may come unstuck if there is a very heavy fall of powdery snow which covers all the seeds. In such circumstances the Bearded Reedling population may drop by 90 per cent or even more over the course of a single winter.

These two options may seem to be quite attractive when the alternative course of action is to migrate thousands of kilometres to a completely different area. Indeed, in human terms, the thought of having to travel such distances without external aids seems an almost impossible task. Even the nomadic peoples of the Arctic and of the world's great deserts did not regularly journey more than a few hundred kilometres each year. It is a very different proposition for a bird. The amount of effort expended in flying is nothing like as much as a man expends in walking the same distance. The most aerial species, such as the Common Swift, may well fly over a thousand kilometres in a day, even during the breeding season when it is only flying to feed and not to migrate. The metabolism of migrant birds is also very different from that of a human athlete since they are able to mobilize reservoirs of stored fat over short periods as fuel for their long flights. To the birds, as they have evolved through millions of generations, flying comes easy and the option to migrate is the one that very many choose. Once more, thinking of the migration option in human terms the risks may seem enormous. Can the bird store the energy needed to make the flight? Can it navigate correctly to reach its destination? Will there be the proper food supply available when it gets there? Will it be able to make the return trip to the breeding grounds for the following summer? So many hazards still seem to put the migrant species at a disadvantage compared with one which is able

Top Members of the tube-noses, such as this Fulmar, are probably the most completely adapted bird group for a life at sea. They use their highly efficient flight behaviour to effortlessly travel vast distances.

Above Shearwaters, such at these Sooty Shearwaters, are superb ocean navigators and often have transequatorial routes that lead them over many thousands of kilometres of open sea.

For many waterfowl species, that are dispersed over wide areas of arctic or sub-arctic wilderness in the breeding season, winter is a time of crowding together in huge flocks. These are Snow and White-fronted Geese in California at Tule Lake and below are ducks in New Mexico at Bosque del Apache. Concentrations of thousands of birds are quite frequent.

to scrape an existence at 'home' during the winter. An example from Britain of two closely related species shows that this is not the case. Two British researchers, Rob Fuller and David Glue, analysed the nest record data gathered by the British Trust for Ornithology on Stonechats and Whinchats. These species are so similar that many bird-watchers have difficulty in distinguishing the young birds and females. However, the Stonechat is only a short-distance migrant in Britain – a few move as far as Spain but many remain at coastal localities within Britain. The Whinchat, on the other hand, is a full scale trans-Saharan migrant with the British birds probably wintering south of 15°N in west Africa. Both are insectivorous birds of heathland. Since the Stonechats appear on the breeding grounds (if they left them at all) in February they are able to have three broods during the summer as compared with the maximum of two that the Whinchat can manage. The clutch size of the Whinchat is slightly larger than that of the Stonechat but its breeding success is markedly worse – half as many nests started produced young as for Stonechat. Overall each breeding pair of Stonechats produces, on average, more than two and a half times as many young as each pair of Whinchats. However, Whinchat populations are perfectly healthy and the species can obviously thrive with a much lower productivity than its sibling. Stonechat populations are subject to violent fluctuations after cold winters in Britain and the rest of western Europe. Their higher productivity allows for such losses to be made up quite quickly but they are, nonetheless, much more vulnerable than their migrant cousins. With this species pair the risks involved in long-distance migration are nothing like so heavy as those involved in staying within Europe for the winter.

The migrants

Later chapters discuss specific aspects of migration in detail, often using particular species as examples to illustrate the points being made. However, some very well known migrants may be missing from these examples and so it seems appropriate to introduce them,

group by group, before going into details. Where appropriate the alternate American or English name for a species which occurs in both regions is given in brackets.

Divers (loons in North America) and grebes. Divers generally move south to coastal waters for the winter. Some grebes are resident, others move south: Horned (Slavonian), Red-necked, Western and Great Crested may often be found in coastal waters.

Albatrosses, petrels and storm petrels. These tube-nosed seabirds include species that move enormous distances both on migration and during their normal feeding flights. The greatest migrants are the truly pelagic (ocean-going) species. However, a few remain in the vicinity of their breeding colonies throughout their lives in tropical and sub-tropical areas. This group includes most of the southern hemisphere breeding birds which regularly venture north of the tropics.

Pelicans, gannets and cormorants. Many of these birds retreat from the northern part of their range during the winter. Flocks of migrating pelicans, cormorants or even anhingas often make use of rising air masses during their migration. The North Atlantic Gannet can be found far out to sea and very much further south than the breeding colonies during the winter.

Herons and storks. Some populations of herons are virtually sedentary but almost all species are strongly migratory in at least part of their range. The same is true of the ibises, bitterns and night-herons. The storks include the European White Stork, one of the best known of all migrants and the subject of many pioneering studies.

Ducks, geese and swans. Although many of the species are widely distributed in temperate areas the migrations of ducks are both complex and unexpectedly extensive. In North America the flyways are mainly oriented north-south but, in the Old World, huge numbers of ducks move westwards out of Russia every winter to western Europe, the Mediterranean and even Africa. Arctic breeding geese and swans follow traditional routes to particular wintering areas. The migrations of this group have been studied intensively and internationally for many of them are favourite quarry species traditionally exploited by sportsmen in the winter far from their breeding grounds. With many species birds from sedentary stocks may pair with migratory mates and move with them to unfamiliar areas – the phenomenon of abmigration.

Raptors – diurnal birds of prey. Some of the most spectacular sights in ornithology are the massed flights of birds of prey as they migrate along traditional routes which make the best possible use of the local topography. These may be, as at Hawk Mountain in Pennsylvania, along a mountain ridge or, as at Falsterbö in Sweden or the Bosphorus in Turkey, where the birds are able to make the shortest water-crossing possible. At such places buzzards, hawks, kites, harriers and eagles are well represented as they depend on soaring during their migration. The falcons also migrate but they are not so dependent on rising air as the other groups. The Osprey is generally a long-distance migrant both in the Old and New World.

Gamebirds. Most species are either sedentary, nomadic or trickle south during the winter months over short distances. There is one startling and marked exception in the European Quail which is a regular, long-distance migrant in enormous numbers.

Cranes. Most populations within our area are migratory moving along a traditional route to traditional wintering areas. This conservatism makes them vulnerable to man-made changes of habitat and several species are severely threatened.

Crakes and rails. Many of the species are insignificant skulking birds of tangled vegetation which appear to be weak flyers. However, the majority of the species are long-distance migrants; for instance, the Corncrake, a regular trans-Saharan migrant, has, in the past, been an accidental vagrant to North American when its populations in Europe were at a high level!

Waders (known as **shorebirds** in North America). The waders include many very long-distance migrants which may even move from breeding grounds north of the Arctic Circle to winter south of the tropics. However, as with the ducks, their movements are very complicated with some extensive east-west journeys superimposed on the north-south ones of other species. As with the cranes, traditional localities are often regularly visited. Many North American species have been recorded in Europe and several European ones are regularly found on the east coast of North America. Their swift, direct flight, often at very high altitudes, may carry them for long distances over totally unsuitable areas from which they are seldom recorded.

Skuas or jaegers. These are coastal birds which breed at high latitudes (including the Antarctic), sometimes inland, and which roam over almost all the world's oceans during the winter. Skuas from the Antarctic are, with the tube-noses, the only regular migrants from the south reaching north of the tropical zone. On the breeding grounds some species, particularly the Pomarine and Long-tailed, may be nomadic from year to year settling to breed only where they find a good food supply.

Gulls. A few species are extensive travellers – like the Sabine's Gull and the Black-legged Kittiwake – regularly covering long distances and being found far out to sea. The majority of others at least wander over moderate distances and some move extensively, but generally from inland (or coastal) breeding grounds to warmer coastal areas for the winter. Patterns of movement in the really successful and recently increased species, like the Herring Gull, are in a state of flux.

Terns and skimmers. The most northern species are all migratory and travel great distances. Indeed the most travelled species in the world is the Arctic Tern which breeds far north into polar regions and winters off the Antarctic pack ice. Many other species reach the tropics and cross the equator but some populations in southern regions remain in the vicinity of their breeding colonies throughout the year.

Auks. Virtually all the species are migratory in at least part of their range. Their movements are often related directly to fish stocks or other food supplies and their autumn (and even winter) movements may be directed northwards rather than southwards from some populations. The birds which breed furthest north are very much constrained by the extent of the ice during the spring and may not be able to reach their breeding grounds in time to raise young in years when the ice does not break-up until very late.

Pigeons and doves. Local movements, particularly the gathering together of large flocks in the autumn, often encourage the belief that even the populations that are relatively sedentary move a great deal. This is particularly the case, as with the Woodpigeon in Britain, when the birds cause a great deal of damage to crops. In both the New and Old World some do move considerable distances and the Turtle Dove in Europe and western Asia is definitely a summer migrant which arrives late in the spring and departs early in the autumn.

Cuckoos. The true cuckoos are all long-distance migrants at least reaching the tropics from the northern temperate regions. In North America the Roadrunner and anis are extending their range but these birds are not migratory.

Owls. Some owls are very sedentary and probably move no more than a few kilometres from their birth-place throughout their lives. Others, probably the majority, can best be described as nomads shifting as their food supply diminishes in one place and recovers in another. Many of these species also tend to wander south during the winter – particularly from areas where snow cover would make it difficult to feed. A few species, notably the Scops Owl which winters in the tropics, are regular long-distance migrants.

Nightjars. Not enough moths remain active throughout the year for the nightjars to be able to over-winter in any but the warmest parts of the Holarctic. Most winter in the tropics.

Above This Sanderling, leaping a wave on a Californian beach, will have been born in the high Arctic – many thousands of kilometres further north.

Opposite top Rows of Cliff Swallows resting on power lines in Oregon in September before their migration southwards.

Opposite below The vast numbers of Little Auks in the colonies of Spitzbergen generally keep in northern waters. However, periodically, fairly large numbers get driven southwards by violent storms so that they may be seen along British coasts and sometimes down into Spain, Portugal and occasionally north Africa.

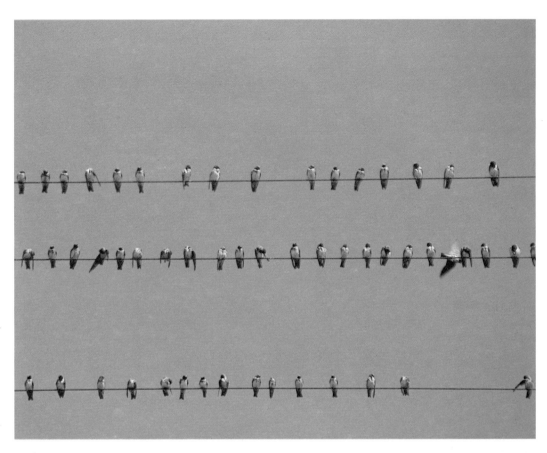

Swifts. As aerial plankton feeders they have no option but to be long-distance migrants from our region during the winter, although some are able to remain throughout the year in the southern parts of it.

Hummingbirds. The only widely distributed hummingbird, the Ruby-throated, over much of North America winters in the tropics. Some of these tiny birds must make a round trip of more than 6000 kilometres from leaving their breeding grounds in the autumn to their return the next spring.

Kingfishers. Their northern limit, during the winter, is dictated by the presence of open water. Some may move east or west to reach areas where the winter freeze-up is less severe.

Bee-eaters, rollers and hoopoes. These insectivorous birds do not penetrate far into the temperate zones of the Old World and are all summer visitors only.

Woodpeckers. Most species are fairly sedentary but a primitive species, the Wryneck, is a full-scale long-distance migrant in the Old World whilst in North America, both the Common Flicker and the Yellow-bellied Sapsucker regularly move quite long distances. None of these three primarily feeds by excavating grubs from wood, the strategy which allows the other woodpeckers to remain in cold areas during the winter.

New World flycatchers. A few winter in southern states but the majority go at least into the tropics and many well into South America. A strongly migratory group of birds.

Larks. Many are sedentary but the most northern species, particularly the Horned Lark of North America and Shorelark of Eurasia (same species), regularly move long distances. Larks are often conspicuous migrants flying together in flocks and during daylight.

Swallows and martins. Since, like the swifts, they feed on aerial insects they are unable to withstand the rigours of a northern winter. For most species all populations withdraw to the tropics or beyond but a few, at the southern edge of our area, are able to remain there throughout the year, examples of which are the Rough-winged Swallow (New World) and the Crag Martin (Old World).

Pipits and wagtails. This mainly Old World group includes some of the insectivorous species of open ground that winter furthest north but also some trans-Saharan migrants, like the Yellow Wagtail and Tree Pipit. The two North American pipits move considerable distances south in the winter.

Shrikes. Some of the Old World species are very long-distance migrants reaching, in the case of the Red-backed species, southern parts of east Africa. The most northern representative of the group, called Great Grey in Europe and Asia, and Northern in America, retreats south in winter to a varying extent from year to year.

Waxwings. These are nomadic wanderers in the winter and may spread far further south in some years if the northern berry crops have failed or their populations are at a high level. May be very tame and common even in suburban gardens where they like to feed on the fruits of ornamental shrubs.

Dippers and wrens. Except where the waters in which they feed freeze, dippers seldom move far. Many species of wrens move out of their northern areas for the winter to warmer parts but they are only medium-distance migrants. Some species are able to survive even in areas of high snowfall by foraging within thickets and clumps of tangled vegetation where the snow does not penetrate.

Mockingbirds, catbirds and thrashers. The few North American species of this neotropical group mostly retreat south in the winter although Mockingbirds may still be found during winter right at the northern fringes of their breeding range in southern Canada.

Accentors. Southern and western populations of Dunnocks are sedentary but those exposed to greater extremes of climate migrate to winter in the Mediterranean area. The Alpine Accentor is generally considered to be sedentary but it does move vertically to winter at or just below the snow-line.

Thrushes and chats. This large and varied group includes all sorts of migrants. The larger species, like the American Robin and the Old World Blackbird and Song Thrush are medium-distance migrants over much of their range but, elsewhere, may be sedentary. Many of the smaller species are long-distance migrants like the Gray-cheeked and Swainson's Thrushes of Canada which winter in South America and Wheatears, Whinchats and Redstarts of the Old World that cross the Sahara into central Africa.

Old World warblers. These include many of the classic European long-distance migrants wintering south of the Sahara in Africa. The Willow Warbler is certainly the most numerous long-distance Old World migrant. In the southern part of our region several species are sedentary. The only North American representatives, the kinglets, like their close relatives the Goldcrest and Firecrest, are medium-distance migrants.

Old World flycatchers. Like the New World flycatchers these are very long-distance migrants. European Spotted Flycatchers regularly reach South Africa.

Titmice, chickadees, creepers, nuthatches and their relatives. This group, with representatives almost everywhere within our region, does not include any long-distance migrants although the northern populations of some species may move south. Several species have populations which may sometimes reach high levels and irrupt moving to areas that they do not normally reach.

Buntings, New World sparrows and cardinals. Many of these species are markedly migratory completely quitting their breeding haunts in the winter. Several of the North American species reach as far south as Panama and some of the Old World buntings winter south of the Sahara. Many species in temperate areas are fairly sedentary.

Tanagers, New World warblers and vireos. The majority of these species are long-distance migrants and many move to South America. However, for about half of the species some may be found wintering in the USA at least in the very southern-most states.

Icterids – New World blackbirds. A few species, for example the Orchard Oriole and Bobolink, are long-distance migrants wintering in South America. The other species only move within North America and some are almost sedentary (except in the northern part of their breeding range).

Finches and weavers. Several of the finch species, particularly Siskins and crossbills, move irregularly over quite long distances – generally in response to fluctuations in food supply. Many other species are regular medium-distance migrants: even the Spanish Sparrow (classified as a weaver) is generally migratory moving south to the arid zone bordering the Sahara in winter.

Starlings and orioles. The Starling is a regular and familiar migrant of continental Europe and central-western Asia; interestingly, the introduced populations in North America seem not to have evolved a regular pattern of migration. The only oriole species to concern us is the spectacular Golden Oriole (the other orioles, of North America, are icterids) which is a long-distance migrant to tropical Africa.

Crows. Most species are sedentary and even northern populations of Ravens may remain throughout the winter on their breeding territories. Some species do migrate from areas which become very cold and others may irrupt irregularly – particularly the jays and nutcrackers.

Above The tiny Ruby-throated Hummingbird is a full-scale migrant and so is able, unlike the other species of hummingbird, to exploit the more northerly and easterly parts of America.

Left Waxwings are irruptive fruit-eating birds which, every few winters, are to be found in flocks much further south than their normal range. These are Bohemian Waxwings – the same species as European ones.

Discovering migration

The historical record of how man came to discover the fact of bird migration and investigate its extent and mechanism is almost as old as history. The skilled bird-catchers and fowlers have, through the centuries, come to understand a great deal about their quarry and, whilst they certainly did not know where they went to, they were perfectly familiar with the regular ebb and flow of bird populations. Such knowledge has been passed down until recent times in areas where communities relied upon seabirds for food, fuel and trade. A good example of this was the lonely island of Hirta in the St Kilda group off western Scotland, until it was evacuated in the 1930s. The tradition of seabird catching is still operating in some areas in the Faroes, Iceland and Greenland.

Archaeologists working in Egypt have uncovered many representations of migrant birds amongst the remains of the old cultures. They include some of the earliest recognizable representations of birds by human artists. Some of the reliefs and paintings date back over 4000 years and show species which were undoubtedly migrants to Egypt even in those days. The identifiable migrants include several species of ducks and three species of geese — White-fronted, Red-breasted and Greylag. These waterfowl were particularly important as food and could only be found in Egypt as winter visitors — except the Greylags which may have bred — the others being high arctic breeding birds. There was no means of preserving fresh meat and so the fowlers would often use their skill to take the birds alive — they would then be kept in captivity and killed when they were needed. This is unfortunate for the Egyptologists who thought, at one stage, that they might be able to date the season of death of the main occupants of some of the lavish tombs by identifying the bird species included as food for the after-life!

Written records of migration are to be found in the Bible and in the writings of several classical Greek authors. In ancient times, as nowadays, some of the most spectacular migration routes of large and conspicuous species like cranes, storks and many raptors skirt the eastern end of the Mediterranean to take advantage of the thermals over the land for soaring. These birds could hardly fail to make an impression. Indeed the birds were familiar to everyone and so, what may well be the first written record of migration – from Jeremiah 8 : 7 – is used as an example to contrast the birds, which know what to do on migration, with the backsliding of the people:

'Yea, the stork in the heaven knoweth her appointed times;
and the turtle and the crane and the swallow observe the time of their coming;
but my people know not the judgement of the Lord.'

The White Stork and the Crane would be conspicuous transients by day. The Turtle Dove and Swallow are not nearly so large but do pass through biblical lands in large numbers at both migrations. The translation of Jeremiah is open to some doubt for, although the references to storks and turtles may definitely be equated with White Storks and Turtle Doves, the other two names may be different species; *agur*, translated as crane, may actually refer to the Wryneck. Cranes are conspicuous migrants through the area – huge and noisy birds of passage – but the much smaller and very skulking Wryneck has a penetrating and characteristic call. Before its recent decline in Britain – from being a familiar bird of the country to a nesting population of a few dozen pairs in less than 50 years – its first calls in the spring meant that summer had arrived and thus was called 'the harbinger of spring'. The other word used, *sus* or *sis*, was translated as Swallow for many years even though ornithologists have known for some time that Swallows may be seen in winter in Palestine. The Arabic for Swift is *sis* and it is now generally accepted that Jeremiah's final reference was to the highly migratory Swift which does not remain in Israel during the winter.

A clear biblical reference to the conspicuous raptor migrations is given in Job 39:26:

'Doth the hawk fly by thy wisdom, and stretch her wings toward the south?'

This is the traditional translation of a verse which has been rendered as:

'Is it of thy devising the hawk grows full fledged, in time to spread her wings for the southwards journey?'

If this version is correct then the author of Job had thought about the very short time that young birds have to consolidate their flying skills before setting out on their migratory flights – an aspect of migration to which many present-day birdwatchers give little attention.

The mention by Jeremiah of the Turtle Dove and the reference to it in the Song of Solomon in the passage celebrating the passing of winter is significant. The traditional sacrifices which were still being practiced in those days were often of birds. The two species most often used were pigeons and Turtle Doves: thus they had a special religious significance. In any case their summer arrival was a happy event:

'For, lo, the winter is past, the rain is over and gone; the flowers appear on the earth;
the time of the singing of the birds is come, and the voice of the turtle is heard in our land;
the fig tree putteth forth her green figs, and the vines with the tender grape give a good smell.'

These references in the Bible have only been observations and not, so far, a matter of life and death. However, migratory birds were then, and in many areas still are, an important source of food. At no time can they have been so welcome and necessary as when the Children of Israel had escaped from Egypt into the wilderness with Moses. Earlier he had purified the bitter waters at the spring at Marah but now the people were becoming hungry. Their needs were fulfilled by the miraculous appearance of Quail and manna. Whatever the manna was, and the arguments as to its identity still continue, the Quail were a very real manifestation of bird migration. The description in Numbers 11: 31 & 32 gives an idea of the amazing event:

'And there went forth a wind from the Lord, and brought quails from the sea, and let them fall by the camp, as it were a day's journey on this side, round about the camp, and as it were two cubits high upon the face of the earth. And the people stood up all that day, and all that night, and all the next day, and they gathered the quails: he that gathered least gathered ten homers: and they spread them all abroad for themselves round the camp.'

It is well known that particular wind conditions can cause, even to this day, spectacular 'falls' of Quail on Mediterranean coasts. The area covered – about 20 kilometres diameter – would not be exceptional but the two cubits depth (about one metre) is obviously fantasy. Rough calculations show that a metre depth of Quails would contain at least 1000 birds per square metre – or 1000 million per square kilometre! However, the biblical story has reached us through several translations and centuries of traditional, oral, story telling. If that reference 'two cubits high upon the face of the earth' referred to the height they were flying it would exactly accord with ornithologists observations of such 'falls'. An eminent ornithologist, Colonel Meinertzhagen, reported seeing such a fall in Port Said, Egypt, when the old men were able to catch the birds as they flew low down the streets at dawn! Surely, in any case, if the birds were so numerous there would have been no need to go out to gather them and certainly not to dry them in the sun for later use.

The Bible is, of course, not primarily a work of natural history and it would be wrong to expect to find a great store of such knowledge in it. The Classical writers are very different for many kinds of writing have survived from Greece and Rome. The histories and plays are the best known but works of agricultural interest and natural histories have also survived. These are generally the gathering together of 'knowledge' by an author for publication, often with little critical appraisal of the merits of the 'knowledge'; indeed, with some writers it is perfectly possible to have totally contradictory facts presented about the same subject in the same book! References to migrants still occurred in other writings in much the same way that Jeremiah used them; for instance Homer likened the routed Trojan army to the fleeing of the cranes from 'the coming winter and sudden rain'.

However, the first naturalist to record his and others' observations, in any systematic manner was Aristotle, who lived and wrote in Greece some 2300 years ago. His books contain references to about 140 identifiable species of birds and in book 8 of his *Historia*

Animalium he deals with migration. His recorded facts and fancies formed the basis of almost all writing on migration for some 2000 years: luckily many of his ideas were sound although their embellishments might be very fanciful. His description of the Crane's migration from Scythia (the steppelands north of the Black Sea) to the marshes at the source of the Nile (Central Africa) is reasonable, although actually it was lifted from the writings of Heroditus about a century earlier:

'There [in the Egyptian marshes] they are said to fight with pygmies. And this is no mere fable, but assuredly there is, as it is said, a dwarf race both of men and horses, and they live in caves, whence they have got the name of Troglodytae, from dwelling in caves. The Grues [Cranes] furthermore do many things with prudence, for they seek for their convenience distant places, and fly high that they may look out far, and, if they shall have seen clouds or a storm, betake themselves to earth, and take rest on the ground. They have a leader also and those who, disposed at each end of the band, may call out, that their voice may be perceived. The others sleep when they alight, with the head hidden underneath the wing, standing alternately on either foot. The leader gazes round him with uncovered head, and by his cry gives notice of whatever he perceives.'

Other classical writers also mention the cranes, not surprisingly since they are exceptionally large birds and regular migrants through Greece. Cicero noted their V-shaped formation whilst migrating and suggested that it was so that the leading bird should keep the wind off the rest. He also observed (and recorded) that the leading bird may change from time to time – as undoubtedly happens – but spoilt this observation by writing that the one that had been relieved was able to rest by placing its head on the back of the bird flying immediately in front! Another Greek, Hesiod, comments that the sound of calling migrant cranes in the autumn was taken as a sign that the rains would soon be coming and that the ploughing should start.

Aristotle's other writings were a similar mixture of facts and theories, which seem quite reasonable to present day ornithologists, interspersed with outrageous embellishments and totally misleading statements.

Opposite top This superb 4000 year old tomb painting shows an Egyptian fowler catching birds. Many of the birds that were caught and eaten were migrants and archaeologists have been able to find the bones of ducks and geese from populations that no longer winter so far south.

Opposite bottom A Short-toed Eagle in flight. This medium-sized predator which feeds on snakes and lizards is a summer migrant to Europe, wintering in tropical Africa.

A cock Wheatear on Skomer island, off the west coast of Wales. The nearby island of Skokholm was the site of one of the first bird observatories in Britain.

Amongst the birds he labelled as migrants were the Pelican (from Bulgaria in the winter to breed on the Danube), Turtle Dove, Swallow, Quail, pigeon, swan and goose. Some of these are summer visitors to Greece and, as we now know, migrate well off the southern edge of the then 'Known World', others are only found in Greece during the winter. He also records that Cuckoos leave early after breeding, at about the time when *Sirius*, the Dog-star, rises: a correct observation as this happens in July. A major discovery, ignored for 2000 years, was his observation: 'All creatures are fatter in migrating'. This fact, actually well known to many bird-catchers who kill and eat migrants, is far and away the earliest reference to pre-migratory weight increase which is practiced by almost all long-distance migrants as a means of storing fuel to power their flights.

The two most unfortunate theories that he propounded were transmutation and hibernation. The former arose because similarly sized and closely related birds were present in Greece, the one for the summer and the other for the winter. Thus Aristotle believed that the summer Redstarts, a common bird through much of Greece, changed to Robins for the winter. The Robin does not breed in Greece but is an abundant winter visitor. The gradual changeover between the two species, over the spring and autumn migrations, would obviously reinforce this idea. He ascribed the change from summer Garden Warblers to winter Blackcaps to the same phenomenon. Indeed he writes that he has seen birds in the half-way stage – between the two species. To us this seems inconceivable but to an uncritical eye a moulting Redstart or Garden Warbler may well seem to be changing into a new and different species – particularly as, with the Redstart, the new winter plumage is nowhere near as bright as the worn male breeding dress which is being lost.

The theory of transmutation lived on for many years as later writers slavishly copied the work of Aristotle. However, as soon as people interested in birds began to travel it became clear that transmutation was not the answer. His other theory, hibernation, took much, much longer to die and, as a supreme irony, almost as soon as it had ceased to be an acceptable theory it became a proven fact for one particular species! Aristotle's writings about other animals, particularly butterflies and small mammals, show that he was quite accustomed to their hibernation in a torpid state over the winter. It was easy to suppose that what they and frogs, snakes and other organisms did to escape the effects of the winter's cold would also be adopted by the birds. He therefore suggested that some birds did not bother to migrate and simply hid themselves in a torpid state in suitable nooks and crannies:

'Swallows, for instance, have often been found in holes, quite denuded of their feathers, and the kite, on its first emergence from its torpidity has been seen to fly from out such hiding-place.'

At first sight these seem to be fanciful and credulous tales which can have no real basis in fact. This is not the case for Swifts nest in holes and their young 'quite denuded of feathers' – incidentally, at that age about the same size as Swallows – can become torpid in cold weather so that they seem very close to death. This is a strategy that they adopt to cope with poor weather conditions when their parents are unable to find enough food for them to carry on growing. Could not such an observation have given rise to the story about the Swallows? Indeed there are a number of records both on spring and autumn migration of Swallows, House Martins and, particularly, Sand Martins being found huddled in clusters in a torpid state when they have been overtaken by inclement weather, even as adults. The kite story also bears closer examination. The returning Black Kites may not be seen arriving from their winter quarters but they will, of course, choose to shelter in a crevice or hole to roost in overnight. When they arrive back, if the weather is cool, they will not leave their roost until the middle of the day when the sun has produced some thermal activity. Thus the first sight of a Black Kite is likely to be one emerging from its overnight roost fairly late in the morning. It would be easy to conclude that this was a hibernating bird just awaking from its winter sleep. Aristotle was not content to stop at the

Swallow and kite when listing birds which might hibernate but included a wide range of other species besides.

The major Roman author on natural history was Pliny whose works are mostly a rewrite of Aristotle's findings of some four centuries earlier. There followed over 1000 years with no recorded advances in the study of migration although there were references to it in some of the literature which has survived. The *Seafarer*, a poem probably written about the Bass Rock in the Firth of Forth, Scotland, towards the end of the seventh century, includes a passage:

'There heard I nought but seething sea,
Ice-cold wave, awhile a song of swan.
There came to charm me gannets' pother
And whimbrels' trills for the laughter of men,
Kittiwake singing instead of mead.'

(Facsimile of the original reproduced below)

All four species are migrants: the wild swans are winter visitors which would be heard whooping as they returned to their breeding grounds in Iceland; the Whimbrels still pass over with their characteristic trilling whistle; and the Gannets and Kittiwakes would be returning to the island for the breeding season. Later the poet also mentions the tern; the translator, James Fisher, from the bird evidence, suggested that the poem had been written in the second half of April.

The major writer on migration, after Aristotle, was the Holy Roman Emperor, Frederick II of Germany. His great book, *De Arte Venandi cum Avibus* (*On the Art of Hunting with Birds*), was written in the first part of the thirteenth century and was mainly concerned with his great love of hunting and falconry. However, these two pursuits require a good knowledge of natural history whilst many subsidiary items are also recorded. He mentions the movements of both the falcons and their quarry. He appreciated that migration was mainly a question, during the autumn, of birds finding warmer regions to spend the winter, a need which could be fulfilled both by travelling long distances southwards and also by mountain-dwelling birds coming to lower levels. He wrote a good deal about the flight formations of geese and Cranes – the most prestigious quarry for the falcons!

For the next few centuries there are some routine references to migration of commoner birds which we know, today, as migrants. One particularly interesting British quotation is from Matthew Paris, writing at St Albans, Hertfordshire in 1251:

'In the course of this year, about the fruit season, there appeared, in the orchards chiefly, some remarkable birds which had never before been seen in England, somewhat larger than larks, which ate the kernel of the fruit and nothing else, whereby the trees were fruitless to the loss of many. The beaks of these birds were crossed, so that by this means they opened the fruit as if with pincers or a knife.'

Top Sandhill Cranes migrating in North America. These magnificent birds make long-distance diurnal migrations from their breeding grounds as far north as Arctic Canada to their wintering areas in southern United States and Mexico.

Above and opposite Ruffs are long-distance migrants which may visit very varied areas in the course of a few months. Here breeding males are portrayed displaying in Sweden (*above*); migrant birds flocking en route in Sudan on the Blue Nile (*opposite top*) and wintering birds, mingling with Lesser Flamingoes and Cape Wigeon, in Kenya (*opposite bottom*).

This is the first reference to a Crossbill invasion affecting Britain. Others were recorded during later years but at long intervals so that the records of the 1593 invasion speak of 'the oldest man living had never heard or reade of any such like bird'. In this case it was generally agreed that they came from some country 'not inhabited' as they took no notice of being shot at and even stayed their ground if apples or stones were thrown at them. The account ends, obviously with relish, by recording: 'they were very good meate'!

Unfortunately, the controversy surrounding the rival theories of migration and hibernation to account for the winter disappearance of birds raged for almost four centuries, fuelled by contributions from many quarters. Aristotle may have started it but convinced adherants of the hibernation hypothesis abounded and produced 'facts' to support their assertions. One such reference, quoted for hundreds of years, derives from *The History of the Northern Nations* by Olaus Magnus, the Archbishop of Uppsala in Sweden. He has the gall to mention that many writers have described the Swallow as migrating to hotter countries for the winter but continues: 'Yet in the northern waters, fishermen oftentimes by chance draw up in their nets an abundance of Swallows, hanging together like a conglomerate mass.' He goes on to say that ignorant young men may carry them to a warm place where 'the swallows, loosened by the access of heat, begin to fly about' but soon perish. The wiser old men put them back without warming them so that no harm comes. The birds get to their wet winter dormitory during the autumn when they assemble in the reeds 'where, allowing themselves to sink into the water, they join bill to bill, wing to wing and foot to foot.'

These stories were believed by many of the best naturalists, even into the nineteenth century. There were also a wealth of additional 'eyewitness' accounts. Linnaeus believed in hibernation of Swallows whilst Gilbert White, although he did believe in migration, had the misfortune to correspond with one of the leading hibernation protagonists – the Hon. Daines Barrington. He was forced to acknowledge, at least to him,

that hibernation of Swallows did take place as the alternative, unthinkable for a gentleman, was to call his friend a liar! In fact even the migration lobby had its lunatic fringe with a pamphlet written 'By a Person of Learning and Piety' and published in London in 1703 in the forefront. In this the moon is suggested as the winter quarters of migrant birds since they are obviously too delicate to undertake sea crossings. It was also suggested that this journey would take up to 60 days, so that the moon had returned to full and was in the same part of the sky that they had set out to reach. Such amazing theories nowadays seem totally ridiculous but it must be remembered that then, 250 years ago, there were few people able to identify birds, the scientific naming of birds was in its infancy and there were great difficulties in travelling outside Europe. Indeed, many of the areas within Africa that we now know are the wintering places of summer visitors to Europe were totally unexplored.

However, there were good observations being made and recorded, at the same time as others were propounding hibernation, as the explanation for the disappearance of the birds in winter. For example, William Turner, later the Dean of Wells, published a book, in Latin, in 1544, on birds which, although it leant rather heavily on Aristotle and Pliny, contained a wealth of new information. For instance, he records the winter flocks of Fieldfares feeding on berries and says it is 'rarely or never is it seen with us in summer'. He also recognized that the Redwing was a winter visitor and a different species from the Song Thrush. His commentaries on what had been written by Aristotle and Pliny demonstrate that he had a very healthy scepticism when it came to the more far fetched of the stories. His debunking passage on the transmutation theory of summer Redstarts turning into winter Robins could hardly be surpassed by a modern ornithologist.

Even before Turner's book, the very first natural history of the Americas was published by Oviedo: *Historia general y natural de las Indias, islas y tierre-firme del mar oceano*. He had seen and recorded the passage of birds in March across the Caribbean and

> They breed in Summer-time in the Fens of *Lincolnſhire* about *Crowland*. They are fatted with white bread and milk, like *Knots*, being ſhut up in cloſe dark rooms: For let in but the light upon them, preſently they fall a fighting, never giving over till one hath killed the other, eſpecially if any body ſtand by. The Fowlers when they ſee them intent upon fighting, ſpread their Nets over them, and catch them before they be aware.
>
> In the Spring time they come over alſo to the *Low Countries:* And it is reported, that at their firſt coming in the beginning of the Spring there are many more Cocks than Hens, but that they never ceaſe fighting till there be ſo many Cocks killed, as to make the number of both Sexes equal.

northwards through Latin America. In particular the migrating hawks impressed him – so numerous that they covered the sky – but he made the mistake, because he did not see the return autumn passage of believing that they simply carried on and circumnavigated the globe each year. Some twenty years later the French ornithologist, Pierre Belon, in the first publication in French on ornithology puts forward evidence for migration both from personal observation and also from experiment. He was a much-travelled man having been to the Eastern Mediterranean, Asia Minor and North Africa and was able to report on seeing migrants both on the move and also in their winter quarters. He also reports the simple experiment of keeping, in a cage, migrant birds with the conditions in which they are supposed to hibernate. None did.

A later writer, John Ray (1627–1705), in *The Ornithology of Francis Willughby*, tried to bring together all extant knowledge on birds for comprehensive treatment.

John Ray describing the migration of Ruffs in 1678.

His section on fowling records the coming of breeding Ruff in the spring in the Fens of Lincolnshire. It is only in recent times that this species has returned as a breeding bird in Britain.

Another shining light of common sense was produced by Forster in 1808 and was reprinted several times in new editions. Essentially a statement of knowledge, worldwide, on Swallows, Forster describes all the species then known (including, as was normal at that time, the Swifts) and devotes several pages to the debunking of hibernation as the means by which they survive the winter, producing good evidence against it. This includes the experiments conducted on unfortunate Swallows by enquiring naturalists who discovered that they did not survive long when submerged, or voluntarily dig themselves, into the mud of a tiny pool built into their aviary. He also quotes a long line of

A highly fanciful woodcut published in 1555 purporting to illustrate fishermen whose nets have brought in both fish and swallows! The author was Olaus Magnus, Bishop of Uppsala, Sweden.

Above Poor-whills of this species (*Phalaenoptilus nuttallii*) are among the very few birds to hibernate with a lowered body temperature.

Left Wintering Sand Martins and Swallow in Kenya. A great deal of research has been carried out on the migrations of these birds and there is some evidence that different populations of swallows have discreet wintering areas in southern Africa.

Opposite top The beginnings of the scientific questioning of bird migration are incorporated in some of these interesting suggestions for further investigation set out in 1678 by John Ray.

Opposite centre The massive migration of birds of prey through the Bosphorus is a centre of attraction for today's birdwatchers. This is not a modern discovery, for here John Ray writes about it 300 years ago. Incidentally, the English Kites were 'Red' and did stay the whole year through.

Opposite centre John Ray, in *The Ornithology of Francis Willughby*, writing about Fieldfares.

Opposite bottom John Ray, as can be seen in this extract, was sure that Swallows migrated and did not hibernate in ponds.

What in the water, as *Morehens?* What Birds fit always on the ground, never lighting upon trees? What perch upon trees?

18. What Birds hide themfelves or change places, whether in Winter or in Summer?

19. What would become of *Nightingales, Cuckows,* &c. in Winter; and of *Fieldfares,* &c. in Summer, if they were kept in Cages, and carefully tended, fed and cherifhed?

20. How cometh it to pafs that the moft vehement cold in Winter-time, if they have but food enough, doth not congeal or mortifie the tender bodies of fmall birds?

Kites they fay are Birds of paffage, fhifting places according to the feafons of the year. When I was once (faith *Bellonius*) on the fhore of the *Euxine* Sea, on *Thrace*-fide, about the latter end of *April,* on a certain very high Hill, near to that Pillar which is at the mouth of the *Bofphorus,* where a Fowler had fpread Nets for catching of *Sparrow-Hawks,* which came flying from the right fide of the Sea; we obferved *Kites* coming thither in flocks, and that in fo great numbers, that it was a miracle to us. For being as it were aftonifhed at the ftrangenefs of the fpectacle, we could not conceive where fuch a multitude of *Kites* could get themfelves food. For fhould they for but fifteen days fpace fly continually that way in fuch numerous fquadrons, I dare confidently affirm, they would exceed the number of men living upon the Earth. Howbeit, with us in *England* they are feen all the year, neither do they fear or fly our Winters.

Thefe Birds fly in flocks together with *Stares* and *Redwings.* They fhift places according to the feafons of the year. About the beginning of Autumn come over incredible flights of them into *England,* which ftay with us all Winter, and in the Spring fly all back again, not one bird remaining; infomuch that (as far as ever I could hear) there was never feen young *Fieldfare* or *Redwing,* or fo much as a Neft of thofe birds with us in *England.* Whither they betake themfelves, or where they breed is not to us perfectly known: It is by fome reported, that they breed in *Bohemia;* others tell us with much confidence, in *Sweden.* They have a hoarfe chattering note, not much unlike a *Magpie;* by reafon the fides of the fiffure in the Palate are rough; as we conjecture.

What becomes of *Swallows* in Winter time, whether they fly into other Countries, or lie torpid in hollow trees, and the like places, neither are natural Hiftorians agreed, nor indeed can we certainly determine. To us it feems more probable that they fly away into hot Countries, *viz. Egypt, Æthiopia,* &c. then that either they lurk in hollow trees, or holes of Rocks and ancient buildings, or lie in water under the Ice in Northern Countries, as *Olaus Magnus* reports. For as *Herodotus* witneffeth, they abide all the year in *Egypt,* underftand it of thofe that are bred there (faith *Aldrovandus*) for thofe that are bred with us only fly thither to winter. I am affured of my own knowledge (faith *Peter Martyr*) that *Swallows, Kites,* and other Fowl fly over Sea out of *Europe* to *Alexandria* to winter.

naturalists who support the theory of migration which includes Bewick and Pennant. He also records that he has been able to get evidence from many sailors that Swallows – and he is careful to make sure they are not terns (often called sea-swallows) – have been encountered on migration far from land. His conclusions were forthright:

'In fine, the result of my researches on this subject has convinced me, that the swallow is a migratory bird, annually revisiting the same countries, in common with other birds of passage; and that the bulk of each species betake themselves to some warmer climate when they disappear in autumn. There is sufficient evidence on record to establish the migration of birds of this genus; at the same time that, from the inaccurate observation of the witnesses, it is difficult, in most cases, to determine exactly the species alluded to. But while it is pretty certain that the greatest number of swallows migrate, it is not impossible that many individuals of each of the species may be concealed during winter near their summer haunts. Nature may have provided the swallow with this power of accommodating itself to accidental circumstances; and have enabled it, when hatched late, or otherwise prevented from joining the annual emigration, to sleep in security through the season when it could not obtain its proper food abroad; and to be revived again on the return of warm weather and of food.
On the other hand, as there exists no proof of the vernal reanimation of torpid swallows, it is possible that their torpidity, perhaps induced merely by cold and hunger, may, unless they be roused by accident before it has gone on too long, be a fatal period to their existence. The cases of the discovery and revival of such torpid swallows are surely interesting; and future investigations may, perhaps, throw some light on the destiny of those left undisturbed.'

Clearly he realized that the discovery of some torpid birds in the autumn was very different from proof that the same birds were able to remain torpid through the whole winter and then revive.

However, having taken some 2000 years to refute the theory of hibernation, it is now known that there is a bird which does hibernate through the winter and survive. In 1946 a Nuttall's Poorwill was found in a torpid condition, with a very much lower body temperature than normal, hibernating in a crevice in the Colorado Desert in California. This small nightjar was able to survive the winter and was found again for the next three winters in the same crevice. Later, other members of the same species and the related Trilling Nighthawk were induced to hibernate in captivity. In their state of suspended animation only 10 grams of fat were needed to sustain them for 100 days.

The middle and later parts of the nineteenth century saw a much more scientific approach to ornithology with a great number of skins of birds collected from all parts of the world. The ornithologists working on the taxonomy of the birds collected were often able to determine the basic patterns of migration from the season during which killed and distribution of the species from different parts of the world. Very often the naturalists in the field, collecting birds for museums and private collections, would recognize species as only appearing in certain seasons and the literature on migration blossomed. The more modern developments will be dealt with in their appropriate chapters but there are two which occurred in the nineteenth century. The first was the original 'bird observatory'. It was very different in style from the modern observatories which are familiar to many ornithologists now birding in Europe and America. This was the island of Heligoland in the south-eastern North Sea and the work of one man, Heinrich Gätke, who spent more than 50 years on the island birdwatching, making notes and collecting skins. From his work and that of his contemporaries a great deal of information on the timing and extent of migration were discovered. The emphasis on collecting meant, as has happened always with keen birders, that great store was set by the very rare birds that he found. His classic book, *Heligoland as an Ornithological Observatory*, was translated into English and published in 1895. He had recorded, up to 18th April of that year, 398 different species on an island tiny in comparison with Britain where, to this day, only about 25 per cent more species have been recorded. The book is well worth reading for it also shows how little was understood of the mechanics of migration. For instance Gätke wrote chapters in his book about the altitude and velocity of migration. The former, by and large, he got right with estimates of the height of migration reaching 5000 metres and more, although his suggestion that birds regularly reach 10,000 metres is something of an exaggeration. The velocity he over-estimated by a factor of six- to eight-fold; he suggested that a Bluethroat could manage '. . . at least 1600 geographical miles in the space of nine hours'.

The rare birds that Gätke shot on Heligoland were quite a valuable commodity as almost all serious birdwatchers in the midpart of the last century also collected skins and eggs. Indeed there were several families of 'shooters' on the island who provided commercial suppliers of specimen birds with their stock in trade. This situation was by no means unique and in Britain the birdcatchers on the Sussex Downs, whilst mainly concerned with catching Wheatears and larks for the table, also did a good trade in rarer species for the many local taxidermists. However, at the other end of the North Sea, the foundations were being laid for the fame of another island, Fair Isle, situated halfway between Shetland and Orkney. Here ornithologists like Eagle Clarke and the Duchess of Bedford, interested both in migration for its own sake and in the rare birds for which the island became famous, regularly came to collect birds. The local islanders were, as on Heligoland, able to earn extra money by shooting rarities for collectors and this practice continued through the years to the foundation of the Bird Observatory in 1948. This will seem strange to most present day European ornithologists since the days of collecting are long past but, in North America, it is only in recent years that sight records have been acceptable in many states. The difference in outlook may be best encapsulated by the fact that the Duchess of Bedford, at the time that she was a regular collector of skins from Fair Isle, was the President of the Royal Society for the Protection of Birds!

The other major advance in migration study during the nineteenth century was the very start of what would, today, be called 'Network Research': the gathering of information through an extensive network of observers, most of which are amateurs. The earlier writings of Gilbert White and other naturalists, who kept 'nature diaries' noting the dates of arrival and departure of birds, the flights of moths and butterflies, flowering of plants etc., were widely read and their ideas copied. As early as 1834, there was a proposal in Britain, not acted upon, that a chain of coastal observers should be set up to plot the movements of seabirds. Elsewhere in Europe such schemes did start, generally under the auspices of the national 'scientific academy', with each participant being sent questionnaires to be completed and returned to the centre for analysis. National schemes in Belgium, Sweden, Russia and Germany had produced reports by the 1880s.

British efforts were then just starting with a group of naturalists, including Gurney, Cordeaux and Harvie-Brown, sending forms out to some hundred lighthouses round the British and Irish coasts. They had realized the potential of recording migrant birds attracted to the beams of the lights; following the success of their private pilot scheme they approached the British Association for the Advancement of Science and a 'Committee on Migration' was formed. This eventually included not only Professor Newton, the doyen of British ornithology at that time, but also Eagle Clarke and other active ornithologists. This group was disbanded a decade or so later but both Eagle Clarke and other members maintained their interest; its continuance, some 15 years later, can be seen in the special investigation on migration carried on by the British Ornithologist's Club from 1904 until the outbreak of the war.

As early as 1866, Spencer Fullerton Baird summarized the American data on migration. When the American Ornithologists' Union was formed in 1883 one of the earliest committees to be appointed was to investigate migration, with Dr Merriam in the chair. Within two years government money was forthcoming and Dr Merriam became the first director of the Division of Economic Ornithology. With Professor Wells W. Cooke, who succeeded him, the investigations on migration resulted in a steady increase in knowledge, and a stream of publications, spanning 35 years. These were, however, also the years that saw the extinction of the most numerous North American migrant: the Passenger Pigeon. An alternate name for the bird was Migratory Pigeon for it retreated from the colder areas where it bred for the winter and returned again in the early spring. The migrating flocks were immense, even within a few decades of the bird's eventual extinction. The description of one flight by Major King at Fort Mississauga, Ontario gives a flavour of the vanished spectacle:

'Early in the morning I was apprised by my servant that an extraordinary flock of birds was passing over, such as he had never seen before. Hurrying out and ascending the grassy ramparts, I was perfectly amazed to behold the air filled and the sun obscured by millions of pigeons, not hovering about, but darting onwards in a straight line with arrowy flight, in a vast mass a mile or more in breadth, and stretching before and behind as far as the eye could reach.

Swiftly and steadily the column passed over with a rushing sound, and for hours continued in undiminished myriads advancing over the American forests in the eastern horizon, as the myriads that had passed were lost in the western sky.

It was late in the afternoon before any decrease in the mass was perceptible, but they became gradually less dense as the day drew to a close. At sunset the detached flocks bringing up the rear began to settle in the forest on the Lake-road, and in such numbers as to break down branches from the trees.

The duration of this flight being about fourteen hours, viz, from four AM to six PM, the column (allowing a probable velocity of sixty miles an hour, as assumed by Wilson), could not have been less than three hundred miles in length, with an average breadth, as before stated, of one mile.

During the following day and for several days afterwards, they still continued flying over in immense though greatly diminished numbers, broken up into flocks and keeping much lower, possibly being weaker or younger birds.'

Thus the discovery of migration by mankind has a long history. It has not always progressed in a smooth and logical manner and has, for thousands of years, been a subject of wonder and superstition. There are recurring stories – from northern Europe, the east coast of Britain and from the Indians and Eskimos of North America – that small migrants hitch lifts on the backs of larger and more conspicuous birds. In Siberia it is thought that Corncrakes hitch lifts, presumably because their flight seems too weak to carry them long distances. Similarly Goldcrests are supposed to cross the North Sea on the backs of Long-eared Owls. In the Mediterranean it is commonly recounted that this is how the tiny warblers manage to cross large areas of sea and, only 100 years ago, the August edition of *Nature* published a letter about small birds carried to Crete on the backs of Cranes. An immediate response was received from both Hudson's Bay and Montana where the Cree and Crow Indians, respectively, both believed this to be a fact. In the first case it was supposedly a finch or bunting travelling on the backs of geese but the exact identity of the small bird travelling on Sandhill Crane's backs through Montana is not known. The Cree called it *napite-shu-utl* (crane's back) and the correspondent with *Nature* thought it to be a small grebe which was certainly 'big medicine' if it could be captured. The same stories are also told about Whooping Cranes by Cree Indians. It is easy to understand how these stories originated for the small birds arrive at the same time as the cranes, seem too weak to have flown on their own and the cranes, in flight, make sounds very like the twittering of very much smaller species. It is just possible that a very tired small migrant might land on the back of a larger species but there is no doubt that, if it ever happens, it is a very rare occurrence indeed.

Perhaps the best known totally manufactured story relating to a migrant bird is the miraculous change from barnacle to goose, which was said to account for the arrival of the Barnacle Geese in the autumn. It is explained, and refuted, very well by John Ray in his book, *The Ornithology of Francis Willughby*: (*see opposite*)

In this case the advent of the warm-sea barnacles, carried on drifting wood by the Gulf Stream to the west coast of Ireland, happened shortly before the Barnacle Geese arrived for the winter. I think it more than likely that the original explanation for the tradition is to be found in the Catholic ban on eating meat on Fridays and during Lent. By supporting (even proposing) this bizarre

What is reported concerning the rife and original of thefe birds, to wit, that they are bred of rotten wood, for inftance, of the Mafts, Ribs, and Planks of broken Ships half putrified and corrupted, or of certain Palms of trees falling into the Sea, or laftly, of a kind of Sea-fhels, the figures whereof *Lobel, Gerard,* and others have fet forth, may be feen in *Aldrovand, Sennertus* in his *Hypomnemata, Michael Meyerus,* who hath written an entire book concerning the *Tree-fowl,* and many others. But that all thefe ftories are falfe and fabulous I am confidently perfwaded. Neither do there want fufficient arguments to induce the lovers of truth to be of our opinion, and to convince the gainfayers. For in the whole *Genus* of Birds (excepting the *Phœnix* whofe reputed original is without doubt fabulous) there is not any one example of equivocal or fpontaneous generation. Among other Animals indeed the leffer and more imperfect, as for example many Infects and Frogs, are commonly thought either to be of fpontaneous original, or to come of different feeds and principles. But the greater Animals and perfect in their kind, fuch as is among Birds the *Goofe,* no Philo-fopher would ever admit to be in this manner produced. Secondly, thofe fhells in which they affirm thefe Birds to be bred, and to come forth by a ftrange *metamorphofis,* do moft certainly contain an Animal of their own kind, and not tranfmutable into any other thing : Concerning which the Reader may pleafe to confult that curious Naturalift *Fabius Columna.* Thefe fhells we our felves have feen, once at *Venice* growing in great abundance to the Keel of an old Ship ; a fecond time in the *Medi-terranean* Sea, growing to the back of a *Tortoife* we took between *Sicily* and *Malta. Columna* makes this fhell-fifh to be a kind of *Balanus marinus.* Thirdly, that thefe Geefe do lay Eggs after the manner of other Birds, fit on them, and hatch their Young, the *Hollanders* in their Northern Voyages affirm themfelves to have found by ex-perience.

The arrival of the Barnacle Goose in autumn is well explained here by John Ray.

explanation a perfectly good goose was de-clared to be fish and could be eaten by the devout at times when meat was forbidden. Indeed the same sort of myth was current in many different countries and allowed birds to be eaten as fish at times when the eating of flesh was forbidden. The earliest written reference that Edward Armstrong, author of the definitive work *The Folklore of Birds*, was able to find is almost 1000 years old. It is a myth told by a Moslim diplomat from Ireland.

The most obvious references to bird migra-tion in folklore refer to the returning migrant birds in spring being the harbingers of sum-mer. The almost universal favourite, in both the Old and New Worlds, is the Swallow, probably because of its long-standing habit of nesting on buildings. The English saying:

'One swallow does not make a summer' is echoed in almost all countries and can even be traced back to Aristotle. Ritual welcoming of the returning Swallows has been practised in many areas and other returning migrants may also be greeted. The Cuckoo in Europe was also one of the widely awaited birds and many areas kept special days as those when the first Cuckoo should be heard. Indeed it is the Cuckoo that features in the very old song 'Sumer is i cumen in' – written about 1225 probably at Reading Abbey by John of Fonsete and the earliest surviving piece of English secular music:

'*Summer is a coming in loudly sing cuckoo ;*
Groweth seed and bloweth mead and springeth wood anew ;
Sing cuckoo. Ewe bleateth after lamb, loweth after calf the cow,
Bullock starteth, buck reverteth : merry sing cuckoo, cuckoo, cuckoo ;
Well singest thou cuckoo, nor cease thou never now.'

Patterns of migration

It is inevitably the ambition of scientists studying nature to wish to classify and categorize their observations and discover the 'fundamental truth'. Physicists and chemists are able to predict exactly what will happen when two objects collide or two chemicals are mixed but the biologist, working with living things, is seldom, if ever, able to be completely certain as to what will happen. The study of migration is a biological subject and the individual birds which set out on their migrations are living things responding each to the different stimuli to which they are exposed. It is, therefore, a gross simplification to try to describe the pattern of movements even of a single species by a few arrows on a map. What those arrows represent are the annual movements, over a period of years, of thousands, or even millions, of different individuals all of which will have different paths and different timings of their flights. However, even though they do represent such a gross simplification of a very varied and complex whole, such maps can be very helpful in putting across the differences and similarities between the paths of migration of different birds.

In this chapter maps with the ranges of birds in summer and winter and with arrows

indicating movement will be used to show the wide range of different migration strategies used by birds which visit the Holarctic. The arrows may sometimes only be very general indications of the direction of movement or, in the better known areas and with the best studied species, indications of the actual routes taken by birds which have been discovered by mass ringing studies or detailed observations. However, the idea that the existence of a line on a map actually represents the route taken by an individual bird (or worse the whole population) should generally be resisted, as the individual bird's tracks of the majority of species follow very broad fronts. For a few species, which are funnelled by the lie of the land and by their dependence on thermal activity for soaring flight, the lines on the maps can be taken to give a fair representation of the routes taken.

One principle that applies to all the maps is that the bird's route will often take them over areas totally inhospitable to the species. Land-birds make long ocean crossings, water birds traverse the deserts and open country species overfly the forests. In all cases the migrants are only likely to come to ground where they find a habitat which suits them. This is very restricting for the more specialist species which are only able to use a few sites on migration; these sites take on a very special significance and might, with some species, be so important that their loss could cause rapid extinction of the populations using them. The long-distance migrants, because of the wide variety of habitats that they must adapt to, face a very real problem in finding suitable food during both summer and winter. There is no chance that a migrant summering in the arctic will be able to feed during the winter on exactly the same food as in the summer, thousands of kilometres further north. However, the food gathering adaptations that have evolved are generally to do with foraging techniques which are not designed for feeding on a particular species (the Snail-eating Everglade Kite of the Americas is an exception to this rule) but on a species with a particular life-style. Thus the arctic wader that feeds on small worms in a sandy substrate is equally well equipped to take them whether they are in an arctic beach, a British or New England estuary or the sandy shore of an African or South American lagoon.

One aspect of migration that cannot be shown on the maps is the timing of the flights. These are just as important as the routes taken but are not easily represented on the same maps that show the spatial patterns of movement. Later (page 38) some isochronal maps of the first arrivals of spring migrants in Europe and North America will vividly demonstrate how the birds re-occupy the breeding area like an incoming tide. In general the timing of the northern journey in the spring is the crucial one as the birds which arrive first in the breeding quarters, as soon as conditions are correct, will be able to have the choice of breeding territory. If the bird arrives too late, it may have to breed in sub-optimal areas but, of course, if it arrives too early, the food supply may not yet be available. The departure from the breeding area in the autumn may be crucial in the far north but is a fairly leisurely operation for many migrants leaving temperate areas.

The idea of plotting routes as lines on maps is familiar to all. On a local scale it is quite realistic to use two dimensions to represent an area on the globe; however, when dealing with long-distance migration, it must be remembered that a straight line on a map plotted with a familiar projection (such as the mercator projection used for many of the maps in this book) is not the shortest distance between two points on the earth's surface. Long-distance air travel has helped to make people realize that this is the case; for example, aircraft flying directly from London to Houston, Texas regularly fly over Iceland, Greenland and the Great Lakes! This distortion must be kept in mind whilst interpreting maps of migrants that move several thousand kilometres each year.

Finally, before providing details of the patterns of a wide range of migrations, it is necessary to think of the constraints on a bird's movements at different times of the year. During the breeding season the main constraint, and one which must affect all species, is the need to visit the nest regularly. In most species this is accomplished by foraging close to it, returning frequently. However, many species, notably seabirds, have developed breeding strategies which allow one parent to be away for long periods (often days, sometimes even weeks) whilst the other incubates. This enables feeding to be done hundreds or even thousands of kilometres from the nest. When the chick hatches it may be able to look after itself from an early age and thus allow both parents to forage over long distances. With many of the auks, the chick may fledge within days of hatching and swim with its parents to good feeding areas hundreds of kilometres from the breeding site. Outside the breeding period the bird may have no need to be attached to a particular spot. It may be able to shift as its food supply varies or moves. However, there are a number of species which have been studied in detail which do become attached to particular territories when they are not breeding. Some even defend territories whilst on passage! With such species the pattern of individuals' movements will probably be very clear-cut with little if any shifting about in the winter. Others, which follow transient food supplies like fish stocks in the sea or termite swarms on the land are likely to ebb and flow during the winter.

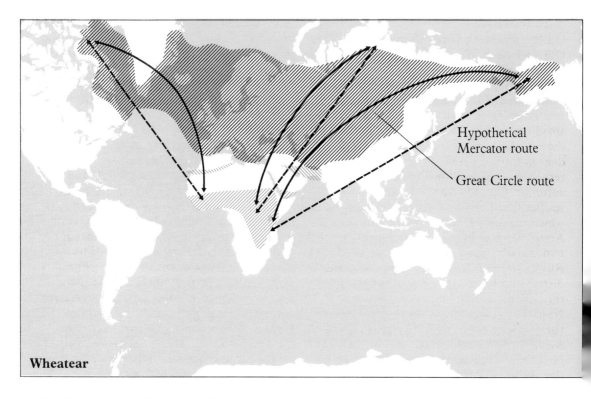

Hypothetical
Mercator route

Great Circle route

Wheatear

The Wheatear is one of the most familiar migrants to birders in Europe as it is one of the earliest summer visitors to pass through in the spring. It is spectacularly widely distributed, with some breeding birds in Alaska and populations in north east Canada and Greenland. All these birds, together with those from the populations breeding across Europe and Asia, winter in Africa. The map above shows these distributions and, with the *curved* lines and probable routes taken by birds from the middle and both extremes of the breeding range to their wintering areas.

Key to maps

On many of the distribution maps hatched areas, indicating breeding or wintering areas of landbirds, may stretch over huge areas and include great tracts of ocean. It is not implied that ships at sea in these areas will have the birds breeding on them but it would actually confuse the map, by concealing the broad sweep of the species' range in tiny detail, if the oceans were left blank and the hatched areas were only printed on the land masses. It would also make it very difficult to show populations on small islands where the hatching would not register. Similarly no attempt has been made to show the detailed distribution of each species within the land areas. For instance, on this map of Wheatears, many parts of southern England should now be shown as blank because the species is only patchily distributed in the intensively cultivated areas.

The straight lines on this map are deliberately misleading. All the maps of distribution (except those of polar distributions) have been drawn on Mercator's projection. This gives a very familiar shape to all countries of the world – except close to the poles where areas are exaggerated – and simplifies the lines of longitude and latitude which circle the earth to a rectangular grid on

The warblers

On the following pages the warblers of the Old and New Worlds provide examples of some of the different patterns of land-bird migrations. The two groups are not very closely related but are ecologically very similar, containing many different species of small, mainly insectivorous birds. None of the American species (the Parulidae) are resident in Europe or Asia although vagrants are recorded crossing the Atlantic in most autumns. On the other hand the Old World species (the Sylviidae) do have three species which regularly breed in North America: the Arctic Warbler in the far north-west and the two *Regulus* species, Golden-crowned and Ruby-crowned Kinglets, which are widely distributed birds moving southwards during the winter.

New World warblers. Rather less than half the 114 species in this family breed north of Mexico. Some are very widespread with members of the same species – for example, the Yellow Warbler – breeding north to Alaska and south to Guatemala whilst others are very rare, like Kirtland's Warbler which breeds only in young jack-pines in northern Michigan. They are all slender little birds between 10 and 20 centimetres long and are mostly to be found during the summer foraging for insects in scrub or woodland.

None are able to remain in the northern part of the continent during the cold of winter and most species move far to the south for the winter. A number are able to stay in southern areas of the United States and a few may even be found in the winter as far north as the Great Lakes. As with almost all migrant species in the New World there are two major geographical features which influence the bird's movements. The first is the twin mountain ranges of the Rockies and the Andes which impose a north-south 'grain' to both continents. The other is the vast obstacle of the Gulf of Mexico and the consequent lack of land at the same latitudes as the Sahara used by European species. The maps show a variety of solutions even within this single group. Some species, like the Orange-crowned Warbler, shun any Gulf crossing and winter along the southern part of the United States and in the northern part of Central America. The Nashville Warbler (not illustrated) moves rather further south but still makes little use of sea crossings. The Wilson's Warbler is more adventurous and, with the Chestnut-sided Warbler, is committed to a trans-Gulf migration. The Mourning Warbler illustrates an extremely well developed use of the western route past and over the Gulf. However, the western route round the Gulf is not the only solution.

the flat paper. The crucial problem, apparent mostly on long distance journeys, is that the shortest distance between two points – clearly a straight line *on the map* – is not the shortest distance between them on the globe. The shortest distances *in real life* are the great circle routes which are on the circle of intersection, on the surface of the earth, of the plane containing the two points and the centre of the earth. These are best represented on the flat map by curved lines like those seen here.

	main breeding area
	marginal breeding area
	main wintering area
	marginal wintering area

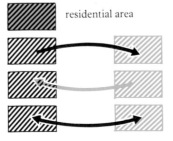

residential area

main autumn migration routes

main spring migration routes

combined autumn and spring migration routes

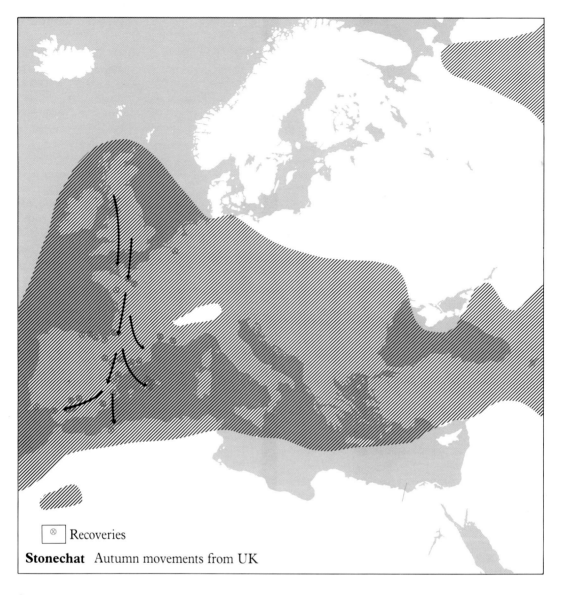

Recoveries

Stonechat Autumn movements from UK

Stonechat

This map shows, shaded, the summer distribution of the Stonechat over Europe. Over much of its breeding area, especially in the south, the Stonechat is sedentary. To the north it is a migrant leaving for the winter but in intermediate areas such as Britain and elsewhere it is a partial migrant. Many British Stonechats move short distances to spend the winter at lower altitudes or on coastal areas where the climate is ameliorated by the influence of the sea. Some are regularly more adventurous and move out for the winter to reach France, Spain and even North Africa. The arrows on the map show the sort of movements made and the points marked indicate where records of Stonechats ringed in Britain have been made.

Such movements are characteristic of short-distance migrants and also of some medium-distance migrants on the southern edge of their breeding range. In all cases where the species maps in this book show an overlap of the breeding and wintering range of a species this sort of system is likely to operate.

A range of species also use the island chain of the West Indies not only as staging posts on migration but also as winter quarters. Many of these species also use the western route, a trend shown by both the Myrtle Warbler and the Yellowthroat (not illustrated). Neither of these species regularly reach South America through the West Indies but many others do. The Northern Waterthrush and American Redstart are classic examples of North American breeding species, very numerous and widespread in summer, which winter almost exclusively in the tropics, reaching the fringes of the Amazon Basin. The most extreme example, and one which will be described in detail later, is the astonishing migration of the Blackpoll Warbler. This species breeds far into western Alaska and yet migrates to northern South America during the autumn by flying out across the Atlantic west of Florida.

Old World warblers. The Old World warblers, the Sylviidae, are a bigger group of species than the Parulidae, with almost 400 species. Many are little known sedentary birds from Africa and Asia but at least 50 are long-distance migrants, including many of the classic trans-Saharan migrants familiar to European ornithologists. Like the Parulidae, they are basically small, slender and insectivorous but with a wider size range from between 9 and 30 centimetres in length – a weight range from about 5 to 60 grams. A number of species are woodland and scrub birds but also included are many which inhabit reed-beds, marshes and rank herb vegetation.

For the long-distance migrants the most important wintering area is undoubtedly Africa, with many species moving westwards from central Asian breeding areas to reach it. Other species also winter in India and the Malay Peninsular but, although several species reach Indonesia, none winter into Australia. From Europe into Africa there are two main routes, one through Iberia and into Africa along the western edge of the Sahara, the other round the eastern end of the Mediterranean and into Africa through the Nile valley. For species moving from Asia, the Himalayan mountain range is a very real

obstacle that tends to bend migration routes east or west around it. Some species only use a single one of these routes but others may use two or three of them. In such cases there is generally a fairly abrupt division between the populations moving one way or another, known as a 'migratory divide'. For instance, the Blackcaps from east of about 10°E in Europe move south-east in the autumn and those to the west of 10°E south-westwards via Iberia to west Africa. Large numbers of Blackcaps also winter north of the Sahara but the majority, at least from the northern part of their breeding range, move further south to winter in central and southern Africa.

This group of warblers, from the genus *Sylvia*, provide other examples of different routes: the Subalpine Warbler (not illustrated) only uses the western route whilst the Lesser Whitethroat, with a much wider breeding distribution, moves only through the eastern route. The even more widely distributed Common Whitethroat uses both routes. Within the leaf-warbler group (*Phylloscopus*), Bonelli's Warbler is mainly a western migrant whilst the Willow Warbler uses both. From Europe the Wood Warbler (not illustrated) is one of the few species which generally flies over the central Saharan region, although minor populations of many other species may do this. Another example of leaf-warblers shows the Arctic Warbler with its wintering grounds in southern Asia. Of the wetland warblers, the Reed Warbler uses both routes to Africa whilst the Marsh Warbler goes round the south-eastern edge of the Sahara and winters in east Africa. As with the other families some species from Asia winter in southern Asia, as shown by the example of the Paddyfield Warbler.

The warblers are widespread and travel widely at night. Indeed, as we shall see later, many of them fly non-stop for several days and nights on end. There are also small migrants that mainly fly during the day – for instance the swallows and martins – and their patterns of movement look quite like some of the maps for warbler movements. However, the means of attaining the same result are very different, with short journeys of a few hours' flight during a day being the rule rather than a few much longer legs. Indeed, a characteristic sight, particularly during the autumn migration, is to see flocks of swallows and martins gathered at night-time roosts or beating southwards over a wide area. Many species are very much less conspicuous on their return passage in the spring when there is such a premium on a fast and efficient journey to allow the breeding season to start at the earliest possible time.

Swifts, swallows and martins

Detailed ringing investigations have provided a mass of information on the autumn movements of Sand Martins in Britain. This small martin (known as the Bank Swallow in America) nests in colonies in sandy faces of quarries or river banks and generally produces two broods each year. In Britain, the youngsters of the first brood are often on the wing by mid-June, almost two months before the peak of autumn migration southwards. These young birds stay in the colonies for a week or so before moving out to form large roosts in reed or osier beds. This movement forms the basis of a gradual investigation of the surrounding area which develops into the first part of the southward migration. In their first few weeks the young birds familiarize themselves with a wide area; during this period they may use several different night-time roosts, even sleeping at active breeding colonies in unoccupied holes. Much of their time is spent in feeding and, presumably, perfecting their feeding techniques but they are always attracted to gatherings of other Sand Martins; thousands of young birds from distant colonies may often visit a large, active colony near a roost. This will help them to decide where to settle when they return from their first migration and, of course, to recognize their surroundings. Gradually they start to move southwards, probably making less than 100 kilometres on any day and moving only once or twice a week.

In southern England, huge numbers may roost in areas where there is good feeding and reed-beds or willows for roosting. On the Ouse Washes, the main drainage channels of the fenland area of East Anglia, it has been estimated that well over a million birds used a single roost in the midst of a large area of cereal crops from which enormous numbers of aphids were flying each day. The main routes at this stage are slightly east of south, probably to ensure the short sea crossing can be made at the eastern end of the English Channel. One morning at another roost, at Chichester in Sussex, I was amazed to watch about half the birds that had spent the night there take off and fly south half an hour before dawn. Because they were favoured by a northerly wind they continued to gain height until they were out of sight and friends watching at Selsey Bill, some nine kilometres further south, did not see them at all – almost certainly the Sand Martins were far too high to be spotted. At this point the Channel is about 120 kilometres across and the marshes of Calvados only half as far again.

Their route is now south-west to the Biscay coast of France and then southwards to travel down the Ebro valley of eastern Spain and along the Mediterranean coast. Some birds cross Spain to the Marismas Marshes, north-east of Gibraltar, before crossing the Straits of Gibraltar and the western edge of the Sahara to winter in West Africa. The maps (see page 60) summarize the movements of the British and Irish populations and also the summer and winter ranges throughout the world of the Sand Martin.

The Swallow (known as the Barn Swallow in North America) is another diurnal migrant travelling relatively short distances on most legs of its migration. Both the species are also forced to be endurance flyers for parts of their migration. For instance, many European Swallows cross the central part of the Sahara where they must certainly fly at least 1500 kilometres with little chance of finding food. Many times birdwatchers have reported totally exhausted birds arriving along the northern edge of the desert in March and April – sometimes dying in vast numbers. There is much evidence for discreet wintering grounds for different populations of Swallows within their extensive winter distribution. The South African ringing of Swallows from the area round Johannesburg has produced two distinct groups of recoveries: the first is basically from Britain and Ireland with a few others from adjacent parts of Europe; the second group is from the Urals of Russia.

The swifts are another group of active and aerial species which show similar migration patterns to the hirundines (swallows). However, some species, like the Common Swift of Eurasia, may not use communal roosts at all as they are in flight from the time they leave their breeding haunts until they return again the next spring. Their migrations are sometimes very fast and they are only in their breeding area for as long as needed to nest and raise their young. The maps show the distribution (see page 61) of the European Swift from Europe and Asia to Africa, the Needle-tailed Swift from Asia to Australia, and the Chimney Swift of the Americas.

Soaring migrants

Although the small insectivorous migrants often travel in flocks and congregate at roosts, their migration is basically on a broad front. The larger diurnal migrants that rely on soaring in thermals for migration have much more concentrated flight paths. Hawk Mountain in Pennsylvania is the most famous site in America with an annual average of 17,000 hawks each autumn and an all-time record day of more than 10,000. There are other sites in America, particularly further south where the birds have been funnelled by the constricting influence of the Gulf of Mexico. In Europe and western Asia there are four main streams where migrant populations from vast areas of the north pass each autumn: across the Straits of Gibraltar, with upwards of 200,000 raptors and storks each year; at the Bosphorus and along the shore of the eastern end of the Caspian Sea in Turkey; and in Israel, at Eilat, where many of the birds from both the Turkish streams pass the head of the Red Sea. These soaring species, which includes not only raptors like the buzzards, Black Kites and Sharp-shinned Hawks but also various storks, cranes and pelicans (even unexpected species like cormorants and anhingas), basically depend on the warming of the air by the sun to create thermals (or kettles) of rising air for migration. This means

that they are unable to migrate if the weather is really bad and cannot start flying, even on a fine day, until the ground has warmed up – say two to three hours after dawn. In an area where raptors are passing through on thermals, like the Gulf coast strip of Texas, grounded raptors may be seen resting in bushes and trees waiting for the thermals to form for them to continue their migration.

Soaring birds have many clues as to where the rising air can be found and over large areas they too are broad-front migrants. However, even relatively short water crossings are shunned; concentrated flights in North America may be found in various places round the Great Lakes and along the indented coasts of the Eastern States and California. In Europe Falsterbö, in Sweden, is another concentration point, in this case for birds leaving Scandinavia. One group in particular has a very strong tradition of visiting the same sites year after year: the cranes. In Europe migrating Common Cranes excite tremendous attention throughout their journey but particularly when they reach Sweden in the spring and can be seen in large flocks before dispersing to their individual breeding grounds. Many cranes are very rare indeed. In North America there are still less than 100 Whooping Cranes alive, an improvement over the minimum of 14 or 15 which constituted the total world population about 40 years ago. In the Old World the rarest are the Manchurian Cranes of Manchuria and Japan. The Japanese population moves short distances but the Manchurian population winters in the northern part of Korea and along the lower reaches of the Yangtse River in China. Another rare Asiatic crane is the Great White Crane whose distributions are mapped (see page 62). The two most extensively distributed Holarctic cranes are the Common Crane of Eurasia and the Sandhill Crane of North America (of which a few also breed in the very eastern part of the Soviet Union). With all the crane species the sight and sound of migrating flocks is unforgettable. In flight they are majestic with trumpeting or bugling calls. On the ground, particularly in the spring, they are superbly graceful creatures and all engage in dancing displays.

Another particularly charismatic migrant in the Old World is the White Stork. All sorts of legends are told about it, the delivery of babies being only a small part of its mythology. Like the cranes, White Storks migrate in flocks and use thermals. They also use traditional routes, particularly in arid areas. Their movements are extensive and have been the subject of very detailed research over many years. The map (see page 64) shows the migratory divide in Europe, as some of the population travels westwards and over the Straits of Gibraltar. The main place for stork watching is the bluffs above the Bosphorus where 200,000 have been counted in a single autumn!

Undoubtedly the most numerous and widespread group of soaring migrants are the broad-winged raptors. Many of the *Buteo* species, the small eagles, the *Accipiters* and even some of the vultures stream southwards during autumn from the northern area and become concentrated at particular places. However, many migrant species have some sedentary populations. Within Europe the western populations of Common Buzzards move relatively short distances; indeed, British and Irish populations seldom move more than 100 kilometres from the nest-site where they were raised. However, the eastern race, *Buteo buteo vulpinus*, moves vast distances wintering as far south in Africa as the Cape of Good Hope. In America the Broad-winged and Swainson's Hawks are amongst the most spectacular raptor migrants, with enormous concentrations making their way through the narrow lands of Central America. One of the most interesting sorts of movement is shown by the Rough-legged Buzzard (known as the Rough-legged Hawk in America) which is the northernmost breeding member of its group. This species depends to a very large extent on the arctic small mammal populations that vary in density in cycles. When populations of small mammals decrease after reaching a peak level the birds migrate much further southwards in the winter than when the mammals are at low density. In such winters they can be found up to 1000 kilometres south of their normal wintering range, regularly reaching Britain and even Ireland in some numbers. The Hen Harrier or Marsh Hawk, another raptor, also spreads further south when the small mammals on which it lives suffer a 'crash' in their population; this also happens with many species of owl.

Not all migrant raptors soar all the time. Some, like the Black Kite (*Milvus migrans*), may use thermals but are capable of flying considerable distances under their own power whilst others, like the Osprey or the falcons, regularly make long-distance overseas flights. The distances travelled by these species are also huge but the true status of the migrants, in the case of the Black Kite and Osprey is often obscured by the presence of resident races in areas where the northern populations migrate through or winter. The Honey Buzzard is also capable of making longer direct water crossings. The small falcon of the north, known as the Merlin in Europe and the Pigeon Hawk in America, has a migration pattern similar to the Rough-legged Buzzard or Hen Harrier but some falcons are very long-distance migrants. The Eleonora's Falcon of the Mediterranean islands winters on Madagascar and in East Africa and is a very late migrant since its breeding season is timed to raise the young on small migrants on their autumn passage. Another falcon, the Red-footed, displays one of the most intriguing patterns of movement. The population breeds in eastern China then undertakes a regular long-distance sea-crossing between northern India and the Horn of Africa, wintering in South Africa. Although no ringing recoveries exist to prove this long-distance movement, the racial differences between the eastern and western populations are unmistakable – indeed the two groups have been treated as separate species by some writers.

Wildfowl migration patterns

The species so far discussed – warblers, swallows, birds of prey etc. – do not have anything like so direct an impact on the lives of men as the wildfowl. Each year several million human hunters wait for the appropriate season to try their skill in killing migrant wildfowl for sport and for food. Over most of Europe and America this involves shooting but all sorts of other methods, including snares, traps and decoys have been used to harvest an abundant and tasty source of food. Certainly the patterns of wildfowl migration will have been very well known to the early humans and this knowledge will have been passed on through the generations, as it is from father to son even today. Hunting is also considered a sport and, as such, is an activity where the best quality 'shoots' are prized, sometimes being worth considerable sums of money. As a consequence a great deal of scientific research, at local, national and international levels, has been funded to determine the details of wildfowl biology – particularly productivity and migration patterns.

We have already seen how many of the soaring birds traditionally use the same route year after year – in many cases a choice imposed by the local topography. The wildfowl are also very conservative in their choice of route and even of local flight-path; year after year the dusk and dawn flights of ducks from marsh to lake will follow the same route which is often much longer than the direct path. It used to be thought these routes were 'handed down' from generation to generation but, although this may have a part to play, it is now realized that subtle differences in local topography probably dictate the same solution year after year. In fact virtually all populations of ducks, geese and swans that migrate have traditional routes, stopping off places and wintering areas. These are most obvious for the rarer and more conspicuous species but they also exist for many of the common widespread ones too. The vast changes man has made to the environment have altered the possible places for special species such as the geese to winter. However, in many cases they have proved adaptable and are able to use agricultural land which has been reclaimed from the wetlands. Many species have actually taken advantage of man-made changes and now reservoirs and drainage areas form the usual wintering places for ducks over much of the temperate Northern Hemisphere.

Wildfowl biologists in North America began to realize, from banding returns and from observations of migrating wildfowl, that there were certain great, traditional routes down the continent in the fall and back again in spring. In 1952 F. C. Lincoln named these routes 'flyways' and defined the four North American ones (see the map on page 71). The term 'flyway' applies equally to the summer breeding area in the north, the wintering area in the south and the network of intermediate routes. A similar system of flyways has been proposed for Eurasian wildfowl migration but it is not so easily recognizable, as differences in geography – mainly the east-west 'grain' of the Old World mountain ranges – tend to blur its existence. The large numbers of national boundaries that the birds fly over also confuses the picture and, until recently with the establishment of such bodies as the International Wildfowl Research Bureau, has greatly hindered concerted research.

The largest wildfowl species are the swans. Even the Mute Swan, virtually a tame bird over much of western Europe, is a migrant having to quit the northern parts of its breeding range when winter produces a complete freeze-up. As a truly wild species, breeding in pockets over the wooded steppe-lands of the southern Soviet Union, the small population moves southwards considerable distances in the winter to reach the southern Caspian Sea and the Iranian coastal wetlands. The high arctic species, known as the Bewick's Swan in Eurasia and the Whistling Swan in America, is a long-distance migrant to traditional wintering areas. Of the two other species, the Trumpeter, in western America, is now much reduced in its population but the Whooper, in Eurasia, is in better shape with over 40,000 birds wintering in Europe.

The majestic sight and thrilling sound of a V-shaped formation of migrant swans is much rarer than the sight of migrating geese. There are literally millions of migrant geese in the northern hemisphere but only about 200,000 of the three truly migrant wild swans. The sight and sound of migrant geese is, however, denied to many birdwatchers who live away from the traditional routes; for them migrant geese may be the most notable manifestation of a particularly cold winter when birds have been frozen out of their traditional haunts.

Even birders used to seeing them regularly cannot fail to be thrilled by the first flocks of the autumn in the south, or the vanguard of the returning multitude in the north. One of the widest-distributed of the basically grey-coloured geese is the White-fronted Goose with populations all round the arctic regions. The other grey goose species are basically Old World in their distribution and mostly move in the same sort of direction as the White-fronts.

In the New World, the dark-necked Canada Geese, in their many different forms, are familiar over most areas where there is suitable wetland habitat. This species has been introduced into many areas and is now common in parts of Europe but, in its native areas, it is one of the classic long-distance migrants. Also in North America are the Snow, Blue and Ross's Goose whose populations range from the high arctic breeding grounds to winter as far south as northern Mexico. The two inland Eurasian 'Branta' geese are much more restricted than the Canadas of North America. The Barnacle Geese breed in three populations each with its own wintering area whilst the mysterious Red-breasted Goose of the Taymyr area of Siberia moves south and west to winter in western Asia, Greece and Bulgaria. This species is decreasing swiftly, giving grave cause for concern. On the breeding grounds it has a nesting association with birds of prey (like Peregrines) and it may be that the severe decline in the raptors has had a knock-on effect. The other Branta species, known as Brent in Europe and Brant in America, is a high arctic breeder mainly wintering in coastal localities.

The movements of the larger wildfowl species have been covered in detail since they are such spectacular birds and are often seen on migration. As a complete contrast the much more common ducks often seem to arrive and depart, on migration, with little fuss and are often unnoticed by the local birders. This is largely because in many areas most species are familiar throughout the year; migrant flocks that are seen tend to be dismissed as local birds moving short distances. Both the dabbling and the diving ducks seen inland do move very long distances, as can be seen from the maps (page 80). However, in areas where the waters do not freeze completely in the winter local birds may remain all winter to be joined by their brethren from thousands of kilometres away. There is an interesting consequence to this mingling since several species pair before moving back to the summer quarters; thus it may happen that young birds bred within the sedentary populations pair with birds from far away and then themselves migrate. This phenomenon, called abmigration, is one spectacular form of gene flow within bird populations and may be responsible for the homogeneity of duck populations within each continent. In general the movements of the sea-ducks are more restricted since salt water does not freeze so readily as fresh water. Most of the eider species move less far than the scoters and both groups will readily make use of large bodies of inland waters like the Great Lakes in North America and the Caspian and Black Seas in Eurasia.

The waders

Just as many of the wildfowl make long-distance movements, from arctic breeding grounds to more equable climates for the winter, so do many of the waders. Some of the longest distance migrants are to be found in this group which includes many trans-equatorial species. Another similarity between the waders and wildfowl is the way that many movements are masked because local populations of the same species are present. In Britain, for example, this masks the midsummer movements of continental Lapwings which journey westwards as soon as they have finished breeding. Indeed, by mid June there are a number of same-year ringing recoveries of young birds which have crossed the North Sea. It is not the familiar waders that provide the most interest for birders; the exotic species which regularly visit ponds, marshes and sewage farms miles from their normal coastal habitat are the most sought-after birds of the true afficionado. One of the first ornithologists to study this phenomenon in detail was David Lack as an undergraduate at Cambridge University in the late 1920s. He and his friends regularly cycled the three miles from the city centre to the old sewage farm at Milton. Here the town's sewage was treated and there were large marshy areas of settlement tanks, alas now all mechanized. More than 30 different species of shorebirds were seen over the first five years of observations.

This proved that many species of waders previously considered to be exclusively birds of the coastal and estuarine mud-flats actually migrated overland. It has since been realized that vast distances are covered overland and across open seas on each migration by many of these species. It has also become apparent that migrant birds, like Knot, which may have travelled several thousand miles to reach their wintering area, will regularly move from estuary to estuary during the course of the winter. Radar observations show such crossings taking place at night from east to west coast in northern England. The overall pattern of Knot movements is, therefore, very much more complex than that shown on the map (see page 81). The map also conceals the species' total reliance on rather small strips of coastal mudflats. One of the most important aspects of international bird conservation is to ensure that such crucial and vulnerable areas are maintained in a fit state to support their bird populations at all times of the year.

Another wader with a very wide distribution is the Sanderling. Like the Knot it is a high arctic breeder but, whilst on passage and during the winter, it is a bird of the open shallow-shelving shore taking tiny morsels of food from the edges of the waves. As the map (see page 81) shows it is rather more widely distributed than the Knot, although the area utilized is still minute compared with a widely distributed land-bird. Not all waders are shore-dwellers, even in the winter. The Old World Ruff (the female is called a Reeve) is a bird of freshwater marshes where the spectacularly plumaged males perform communal 'lek' displays in front of the drab females. During the winter they also live in marshlands but some populations move vast distances. One ringed at the Cambridge sewage farm, some 30 years after David

Lack's work there, was later found more than a third of the way eastwards round the world in Siberia! In North America the Golden Plovers take a trans-oceanic route to reach their wintering grounds in the Pampas of central South America, whilst other populations travel across the Pacific southwards to Australia, wintering on many of the Pacific islands. One of the most famous oceanic-wintering waders is the Bristle-thighed Curlew, actually named after the islands where it was first discovered *Numenius tahitiensis*, which only winters on the islands of the Pacific. The complex and varied movements of waders could take up several books by themselves. Perhaps a suitable example to end with would be the movements of the Ringed and Semi-palmated Plover. These two species, between them, breed right round the arctic with a great deal of sub-specific variation. However, as the map shows (see page 83), in North America the Semi-palmated Plover is a breeding bird of the north which winters along the shores of the Gulf and into South America. In Europe some of the south-western populations are sedentary whilst the high arctic birds overfly them to reach the southern tip of Africa.

Seabirds

The seabirds include some of the world's greatest migrants. Many of the species have a very long pre-breeding period and it is not easy to depict their distribution on maps as, even during the breeding season, non-breeding sections of the population may be a very long way from the colonies. Indeed other birds, with active nests, may often feed hundreds of miles from their nests.

The tube-noses (albatrosses, shearwaters, petrels, fulmars and storm petrels) are probably the most completely adapted group for their life at sea and many never come to land, except during the breeding season. The northern hemisphere has a few native albatrosses in the Pacific (none in the Atlantic) and their movements, although extensive, are more restricted than the wanderings of some of the southern species.

In the Atlantic the largest shearwater, the Great, breeding in vast numbers on the

47

islands of Tristan and Gough in the south undertakes a very long circular migration northwards to our area. In the Pacific the Slender-billed Shearwater, breeding on the islands off south-east Australia and Tasmania, undertakes a similar migration round the Pacific. Such extensive movements are not confined to the southern hemisphere shearwaters. The races of Common Shearwater from Europe – Manx and Balearic – are trans-equatorial migrants moving each autumn southwards to winter in large numbers off the eastern coast of South America. A few get caught up in the westerly winds of the roaring forties and the species has been recorded on the other side of the world – one ringed in Wales was found in Australia. The high arctic Fulmar, found as far north as the edge of the ice packs in most polar areas, moves long distances in the northern parts of the Atlantic and Pacific. With this species the presence of different colour forms, particularly dark ones in northern populations, shows that there is relatively little mixing between north and south populations – although southern birds may venture north. Even the tiny storm-petrels range long distances outside the breeding season. The map (see page 87) shows the range of two species: the European moving thousands of miles south to the tip of Africa during the winter; and the Least, from Baja in California, using the seas along the west coast of Central America south to northern Peru.

The movements of most other groups of seabirds are not so staggering as the tubenoses. Some, like most of the cormorants (a wide-ranging group), never move more than a few hundred miles with the seasons and many populations are quite sedentary. A close relative of the cormorants, the Northern Gannet, is a distant migrant on both sides of the North Atlantic. Some birds penetrate the Mediterranean on the eastern seaboard and the Caribbean on the west. The movements of most auk species are similar to those shown by the Gannet but on a smaller scale. This is shown by the Razorbill which has populations breeding much further north than the Gannets and does not go so far south in the winter. Some of the larger auks are almost

sedentary but the smaller species, like the puffins, become truly pelagic in the winter and roam the open ocean, like the Tufted Puffin of the northern Pacific. The Razorbills and guillemots tend to be birds of coastal waters and continental shelf areas rather than the open sea, since they depend on fish for food, but the smaller species, like the puffins, are able to live on plankton. Outside the breeding season two groups of birds can be thought of as being 'seabirds' although they breed inland on freshwater areas. These are the divers and some of the grebes. Their movements, typified by the Great Northern Diver and the Red-necked Grebe, show a concentration of birds from extensive (but scattered) breeding areas to the coastal waters where they are able to fish during the winter. Both species are often found on inland lakes and reservoirs where their overland migration routes lie.

The final group of seabirds – the gulls, terns and skuas – are a very mixed bunch. Some gulls may be seen almost everywhere in the northern hemisphere, often spending all their lives far from any oceans. Others, like the Lesser Black-backed Gull of Europe and Bonapartes Gull of North America, undertake extensive movements to a mainly coastal existence during the winter. One specialist pair of gulls, the kittiwakes, do very well as fully pelagic seabirds in the winter. The map (see page 88) shows the very widely distributed Black-legged Kittiwake which breeds in the northern parts of both the Atlantic and the Pacific. The Red-legged species is only found on and around the Aleutian chain in the northern Pacific. Tern movements of the species native to our area are mainly distant migrants with the bulk of the populations wintering in inter-tropical waters. An exception is the Aleutian Tern which seems to stay the whole year in the northern Pacific. The most notable wanderer of this group is the Arctic Tern which may well be the world's record migrant bird. Certainly the individuals that breed north of the Arctic circle and winter off the Antarctic pack-ice will experience more daylight during their year than any other living creature. To a certain extent the movements of the Arctic Skuas rely on the terns since they often feed by pirating food from the terns. However, the skuas do not penetrate as far south as the Arctic Terns. This is because there are other species of terns in the areas where they winter and also because they are well able to feed for themselves, without robbing terns, if they need to. The movements of the other three species of skua breeding in the northern hemisphere are also very wide and there have now been several records of southern skuas, from the Antarctic, in the northern parts of the Atlantic.

The passerine migrants

So far the patterns of migration of a number of large but diverse groups of birds have been described. Many other groups include migrants which move north-south in the same patterns as have already been illustrated. For example, among non-passerine land-birds, the Yellow-billed Cuckoo from southern North America winters in South America whilst the Common Nightjar, from Europe and Asia, winters over much of central and eastern Africa as well as in parts of Pakistan and India. Other examples include: the Eurasian Turtle Dove whose wintering area is restricted to the Sahel region south of the Sahara in Africa; the Wryneck, from central Europe and Asia, which winters in central Africa and southern Asia; and the Yellow-bellied Sapsucker, from the central parts of North America, which moves to both the southern areas of the United States and parts of Central America and to parts of the western seaboard.

The passerines, of course, include a very wide range of species, many of which are migrants. So far the warblers have been dealt with in detail but there are large numbers of other long-distance nocturnal migrants which annually make the same sort of journeys. Four examples of very different groups of birds are shown here from the Old and New Worlds. The two flycatchers, Arcadian and Spotted, are not closely related; the Arcadian is a relatively modest migrant but the Spotted Flycatcher is unusual amongst the small Eurasian migrants in penetrating to the very south of Africa. The Redstart of Eurasia and

the Swainson's (or Olive-backed) Thrush of North America are related and their migration patterns are similar; in fact, both species breed north to quite high latitudes and migrate to winter in tropical and sub-tropical areas. The Scarlet Tanager and Red-eyed Vireo are members of exclusively American groups of birds. However, the Red-backed Shrike and Red-throated Pipit do have a few relatives in the New World. Note that in Europe the Red-backed Shrikes travel southeast during the autumn, whilst other similar-sized species move south-west. The Red-throated Pipit breeds in high arctic areas of tundra where winter survival would be utterly impossible. As happens within many groups of birds this species overflies the areas inhabited, both in summer and winter, by other representatives of the genus and winters well to the south.

There are also many species with more restricted movements which often result in an incomplete separation of the summer and winter distributions. The species illustrated (see page 99) show some elements of the population moving very long distances but others, along the southern fringe of the breeding population, may travel only short distances or even stay put throughout the year. The Rusty Blackbird of North America and the Redwing of Northern Europe may cover similar sorts of distances. Results from ringing show that Redwings may use different parts of the wintering quarters in subsequent years; there are many ringing records of birds marked in Britain in one winter being found in later winters in Greece, Turkey or even around the Caspian Sea. The Brambling, the classic winter visitor of the European family of finches is a consistent migrant but does not always use the same wintering area. Vast flocks concentrate on areas where food resources are good; one roost in Switzerland is thought to have contained more than 50 million birds. For this species the food is often mast (beech-nuts) from the Beech which generally fruits prolifically only every few years. The New World example, the Savannah Sparrow, is typical of the sort of movement shown by this varied family.

Shorter range movements, often largely within the breeding distribution, also affects a very large number of species. It is naturally very difficult to depict such movements by showing maps of the whole summer and winter ranges of the species and so movements involving the British Isles only have been illustrated (see page 100). The species shown make the typical movements found at the edges of the continental land masses where rather milder weather is found during the winter than inland but also they may move in very different fashions elsewhere within their range. Indeed the British breeding populations of many species are very much less mobile than other nearby ones, for example the Robin. A small proportion of the local Robins move out to the south-south-west during the winter but most stay within a few kilometres of their summer territories (many do not move at all). However, during the late autumn (mainly October), large numbers of passage Robins from parts of Scandi-

Redwings, such as this individual, visit western Europe in large numbers during the winter months from their breeding territories in northern Europe and Russia.

navia reach the east coast of Britain, most continuing their passage through France to winter in the southern half of Iberia. The incomplete migration of the local birds from Britain during the winter is typical of many species and is generally referred to as 'partial migration'. Some birds, for example the Pied Wagtail, move out in the winter, especially in the northern part of their range in Scotland and in the higher, bleaker areas in the rest of Britain. However, with the Pied Wagtail, the majority of the moving birds only go as far as the southern parts of Britain where winter populations are much higher than summer. This species actually takes advantage of man by roosting communally in areas which have been heated, like greenhouses, power station cooling towers and factories.

For the Goldfinch variable proportions of the population leave for the winter although, in all years, some stay behind and some migrate. The differences may be in response to the availability of food in the autumn but, whatever the stimulus, the two strategies being kept alive within the population allow the species to take advantage of mild winters in Britain without being dependent on them. The movements of the Song Thrush seem, at first sight, to be very similar. Usually, a very small proportion of the population leaves in the autumn to winter further south; however, in response to cold conditions during the winter large numbers of the birds left behind will move south and west for distances of up to a few hundred kilometres to escape the effects of freezing conditions. Such weather movements are also carried out by many other migrants but it is most interesting that birds which are normally sedentary are able to respond so quickly in this way.

Britain also acts as the wintering area for many species whose movements are often oriented east-west during the autumn. The areas of origin of five of these, in each case only a very small part of the overall breeding area, are shown in the maps. To many people the inclusion of Water Rail as a migrant will be most surprising. The rails and crakes seem generally to be reluctant to fly and very incompetent when they do. Many are migrants and there are several, including the Old World Corncrake and the American Sora Rail, which move very long distances into Africa and South America respectively.

Amongst the passerines the Starling is probably the most common winter visitor to Britain – the birds coming from a very wide area of northern Europe. The wintering birds form huge roosts often containing a million or more birds for, during the winter, the breeding population from a very wide area is concentrated into a much smaller region. British Starling populations are mobile over short distances but do not perform regular migrations. British Chaffinches are truly sedentary – there are several records of birds remaining within a few hundred metres of their original ringing sites for six, seven, even

ten years. During the winter they are joined by migrants from Scandinavia and northern Germany. The males and females of this species are easily distinguished and have different migration regimes. The adult males are much more likely to remain near the breeding grounds than the females and young males. When they were first named by Linnaeus the specific name *coelebs* (bachelor) was given them as he only saw bachelor flocks of males in the winter. The last map (see page 105) shows the area of origin of Goldcrests reaching Britain for the winter. These tiny birds are subject to wide fluctuations in population for they suffer very badly in cold winters but Britain, with its warmer climate, is a preferred wintering area for many. It used to be thought so unlikely that such tiny birds could make the journey across the North Sea that it was believed that they hitched lifts on the backs of bigger birds!

Further down the migration scale comes the very short vertical migrations made by many species in mountainous areas when they come down the mountains to reach better conditions in the winter. Even such sedentary species as the Ptarmigan and other gamebirds may come down several hundred metres during the winter. But such movements are easily reversed and, with some species, it is only the average living level which changes, for the birds may maintain their territories at higher altitudes during all but the worst conditions. Indeed some herbivorous species may often find it easier to reach food on exposed ridges, where wind has swept the surface clear of snow, than in more sheltered areas. Vertical movements are at their most pronounced in mountainous areas where there is a wide altitudinal range for the birds to exploit within a short distance. The same effect can be seen even over a few hundred metres of height range where conditions can become critical. Two species whose habitat is freshwater streams illustrate this. The Dipper is a bird which is able to feed underwater and therefore able to live even where the surface is frozen provided that it can reach the water. The rapid fluctuations in height of the water means that it almost always can get in and so it is often able to remain on the streams where it breeds. On the other hand the Grey Wagtail depends for its food on insects flying around the stream and tiny animals stranded in its margins, a source of food not available during the winter when the stream freezes. The Grey Wagtail is therefore both an 'altitudinal' and a partial migrant within Britain.

When, in a following chapter, we consider how the migrants are able to get through their complex timetable of annual events the importance of fitting in the annual moult will become clear. Several species actually moult during the course of their migration, mostly by stopping over at a particular point on their autumn journey. There are, however, some species which need to moult where there is a plentiful food supply and this has forced them to make a special migration, even though they are otherwise sedentary. The species that do this are ducks and geese which become completely flightless during the moult. Two examples, both from Britain where the species concerned are otherwise sedentary, are mapped (see page 105). These involve the native British Shelduck visiting the Heligoland Bight off the German coast in very large numbers where they are joined by birds from many other European populations. The only 'moulting flock' in Britain that has so far been discovered is in Somerset. Birds from west coast estuaries have been followed as they flew inland across northern England at the start of their moult migration to the Bight. For a period of six to eight weeks from late July the Heligoland Bight may have over 100,000 birds present. The area is used as a bombing range but no live bombs are used whilst the birds are present! The other moult migration has only developed recently and involves an introduced population of Canada Geese, in the Midlands of England, which now regularly fly to the Beauly Firth in northern Scotland to moult as non-breeders during the summer. This was first noticed with a few dozen birds about 30 years ago; now almost 1000 birds are involved and there is some indication that they are also coming from further south in England.

There has already been passing reference to weather movements of birds. These are generally brought about by cold weather and involve birds which are seeking warmer areas to survive the winter. Many species are able to undergo these 'contingency' movements and most undertake them regularly when they are forced to. In most species the weather movement is generally short compared with the major migration but there are some, as we have seen, that undertake weather movements even though they are almost sedentary otherwise. It is obviously very important that a bird which is to undertake a weather movement should start off before it becomes weakened by a cold spell and so such movements tend to be concentrated into the very first part of a spell of bad weather. All sorts of species may be involved for freezing conditions are of little benefit to many groups of birds: swimming, diving and dabbling species need open water which will freeze; probing species like waders and thrushes, need soft ground which will become hard; and insect and seed-eaters may have their food resources covered by snow or frozen rime. The only species to benefit may be the predators, who are able to exploit the weakened birds, and the carrion-eaters putting to good use the corpses of the birds and animals that have died. In Spain the Lapwing, a species notoriously prone to long-distance cold weather movements, is known as the *Ave Fria* or 'bird of the frost'. Flocks are likely to appear in the winter when the weather is cold further north in Europe and sometimes, later on, the frosts follow the birds into Spain. There are, of course, other weather migrations of a less voluntary nature as birds are driven in front of hurricanes and other violent storms; however, these cannot correctly be treated as migrations since they are totally involuntary on the part of the bird.

Only one regular type of migration remains – the 'irruptive' movements shown irregularly (or regularly) by some species. These are generally responses to population pressures and food supplies and are exhibited by several sorts of species but mainly by predators and by seed-eating birds which exploit food plants that do not fruit equally well each year. For example, the Rough-legged Buzzard and the Short-eared Owl of the high arctic are able to raise very much larger broods as the cycle of small mammal populations reaches its peak. When the crash comes the birds must venture much further south to find food and survive the winter. Thus, immediately after a crash in small mammal populations of the tundra, many more of these birds and other arctic predators are seen south of their normal range. During the next breeding season, when small mammal populations are low, many of the birds do not breed at all but others may settle to try to breed far from their own natal areas.

Populations of small skuas, Long-tailed particularly, are very mobile on the breeding grounds for the same reasons. The very different extent of movement in different years of species such as Waxwings, nutcrackers and grosbeaks is also related to the food supply. If the populations of the birds are high during the autumn and they are unable to find wintering areas within the normal range with sufficient food they will move further on until they are able to find food. Unlike the cycles in movements of predators, which are often tied to a fairly regular cycle in their prey species, the irruptions of these species are not at regular intervals. Even less predictable are the rare irruptions of tits in western Europe; the last happened more than 20 years ago and involved millions of birds from areas round the North Sea invading Britain with very much more movement within the British population than normal. It was probably triggered by a successful breeding season followed by good winter survival and another successful breeding season; the resulting high population of birds proved to be exceptionally mobile.

Finally the complex movements of crossbills need to be dealt with. They are often not true migrations since they do not involve an annual journey 'there and back'. The crossbill species are dependent on the fruit of coniferous trees for virtually all their food. The trees of a particular area are likely to be very fruitful in one year but not in the next. The successful population of crossbills therefore has to take off and find a new area where there will be fruiting trees for food for the next breeding season. Of course, in some years, this may mean only a short distance move to a neighbouring forest but, in other years, the flocks of birds may wander thousands of kilometres looking for suitable trees and can be found in all sorts of areas where they have not been seen in numbers for decades.

To sum up then there are almost as many different patterns of migration as there are species in the northern hemisphere. However, there are particular themes which emerge time after time, although modified by the ecology of each species:

1) Long-distance trans-equatorial migration to South America, Africa and Asia – even a few to Australia.
2) Long-distance migration within the northern hemisphere but mainly oriented north-south.
3) Species which show long-distance north-south migration for their northern populations whilst their southern ones move less far, if at all.
4) Continental to coastal migration, often with much lateral movement (east-west or west-east), particularly of species able to withstand cold conditions but not prolonged freezing.
5) Long-distance movements of seabirds which are often mobile throughout the time they are away from their breeding areas.
6) Species with populations containing migrant and sedentary birds, i.e. true partial migrants.
7) Altitudinal migrants which seek warmer conditions during the winter at lower altitudes.
8) Irruptive species which only move long distances when population pressures, particularly fluctuations in food supplies, force such moves on them.

Even these categories are not rigid and there are many species whose movements can only be characterized by combining two or three of the categories. What is certain is that as more research is undertaken we shall find that the patterns of movements are much more complicated than we now think.

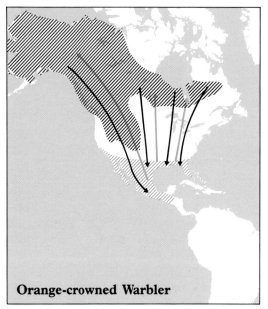

Orange-crowned Warbler

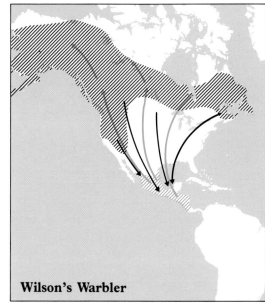

Wilson's Warbler

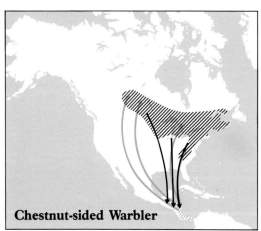

Chestnut-sided Warbler

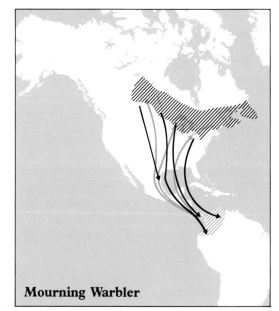

Mourning Warbler

This important group of Nearctic migrants includes a very wide range of species. Many of them are familiar to birders throughout most of North America whilst others are rather local or even very rare. The eight species mapped show some of the patterns of movement and their similarities and differences.

Orange-crowned Warbler
A rather nondescript bird of woodland clearings and scrubby areas. The orange on the crown is hardly ever visible in the field and the bird looks rather like an Old World *Phylloscopus* warbler.

Chestnut-sided Warbler
A distinctive bird with lemon or greenish top to the head, streaked back, white breast and cheeks but a distinctive chestnut line down the sides. Mainly found in open scrublands during the breeding season.

Wilson's Warbler
Another small warbler but the male has a distinctive black skull-cap. Females lack this but both are a bright golden yellow below. They are most likely to be found in damp areas of willow scrub.

Mourning Warbler
This grey headed warbler has olive upperparts and yellow belly and vent. The male has a black breast but the female's is grey. Generally found in fairly thick scrub or understorey.

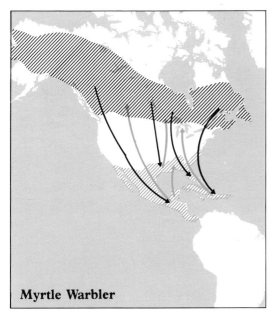

Myrtle Warbler

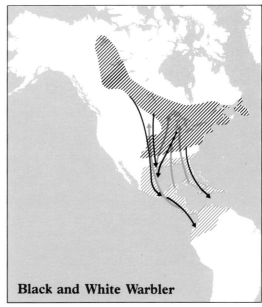

Black and White Warbler

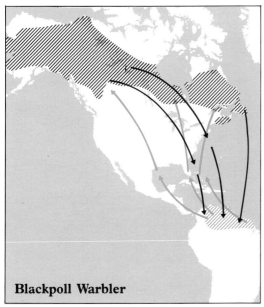

Blackpoll Warbler

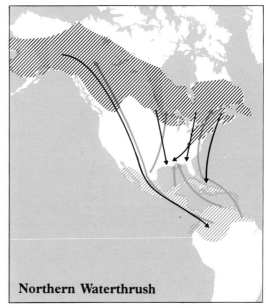

Northern Waterthrush

Myrtle Warbler
This species has yellow on the rump, crown and flanks. The conspecific (but more southwestern in distribution) Audubon's Warbler also has a yellow chin and is *not* included on the map.

Blackpoll Warbler
Superficially like the preceding species but lacks the white crown stripe (head completely black). The male is browner than a female Black and White and the females and immatures hardly have any black on them.

Black and White Warbler
This warbler is perfectly described by its name – it is immediately distinguished from the other two black and white warbler species by the white crown stripe. Much of its time is spent feeding along the trunk and branches of trees like a nuthatch or treecreeper.

Northern Waterthrush
This bird looks like a miniature thrush and behaves like a tiny sandpiper. It is brown above with a pale eyestripe and streaked underparts.

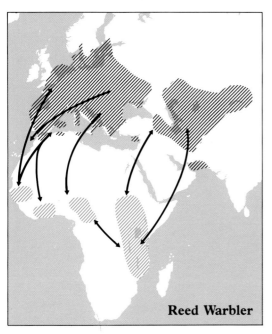

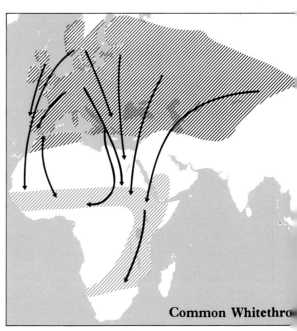

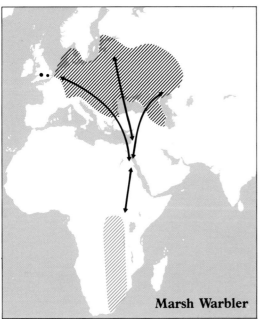

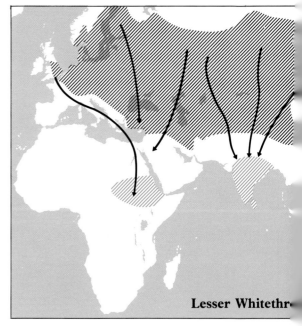

Reed Warbler
A sleek brown warbler of reedbeds or their immediate surroundings building a very characteristic nest supported on last year's dead stems.

Marsh Warbler
A sleek greenish warbler very similar to the Reed. One of the classic bird mimics with individuals producing recognisable snatches of song from 30 or 40 species!

Common Whitethroat
A graceful greyish warbler with a white throat, rusty wings and a long tail. Formerly very common in scrubland and farmland over most of Europe.

Lesser Whitethroat
A smaller and even more attractive version of the previous species. It lacks the rusty wings and has dark lores giving it a cheerfully piratical look.

56

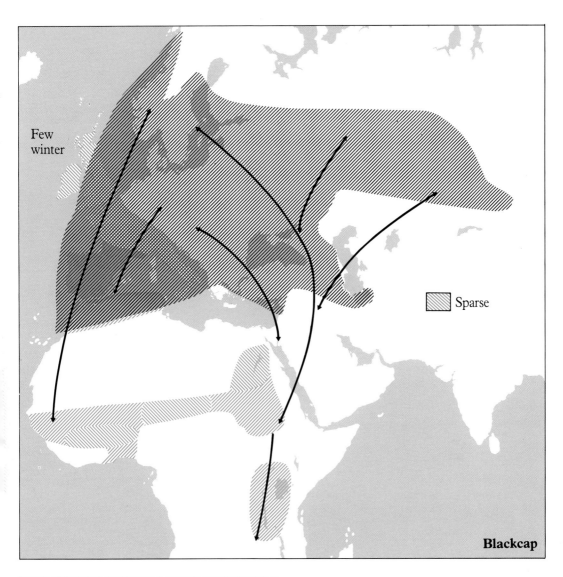

Few
winter

Sparse

Blackcap

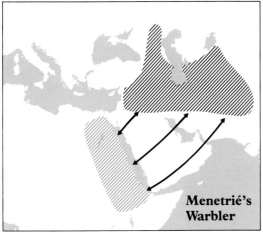

**Menetrié's
Warbler**

Blackcap
A familiar bird over almost all Europe and into Asia.
The male is the sex with the black cap, it is brown in
females and immatures. The complicated system of
movements from their scrub and woodland breeding
areas includes partial migrant populations in the west, a
migratory divide at about 15°E and birds wintering as
far south as the equator in the East.

Menetrie' Warbler
This rather obscure warbler, very like a Lesser Whitethroat but, the males at least, with a rich vinous pink
flush to their breasts has to move out of the scrub and
open woodland areas where it breeds. They would not
be able to survive the cold of a central Asiatic winter.

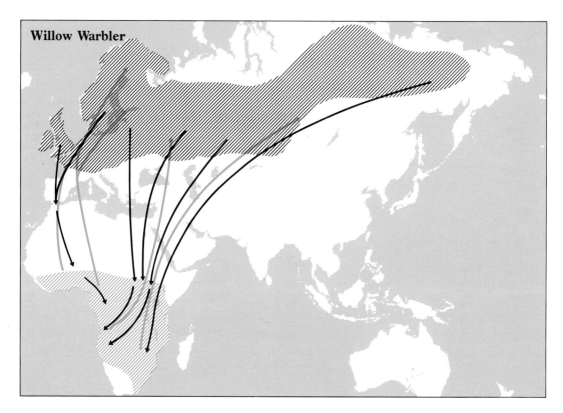

Willow Warbler

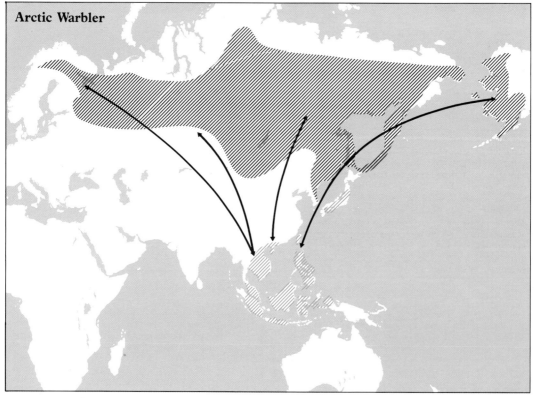

Arctic Warbler

58

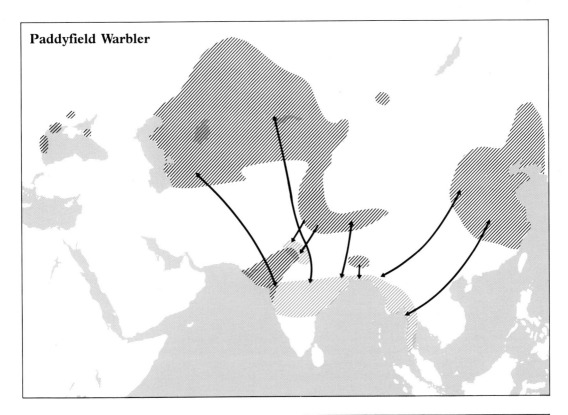

Paddyfield Warbler

Paddyfield Warbler

This species looks very like the Reed Warbler (see page 56) and breeds in reed and grass tangles in wetland areas. Its rather intricate patterns of migration in southern Asia are echoed, at least in part, by other warbler species in the area.

Willow, Arctic and Bonelli's Warbler

These three species are closely related little olive-green birds breeding in open woodland and scrub areas. The first two are very widely distributed across Eurasia – the Arctic breeds also in Alaska – and both perform long-distance movements. Surprisingly they are completely segregated with all the Willow Warblers wintering in Africa and all the Arctics in Southeast Asia. The reason for this can be found in the distant past when the ancestral species was split by the advancing ice into western and eastern species. The Bonelli's is very much a Willow Warbler replacement over southern Europe and undertakes a relatively simple migration across the Sahara. Note that its winter distribution is along the northern edge of the Willow Warbler's in winter.

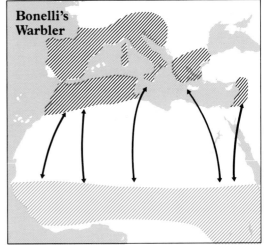

Bonelli's Warbler

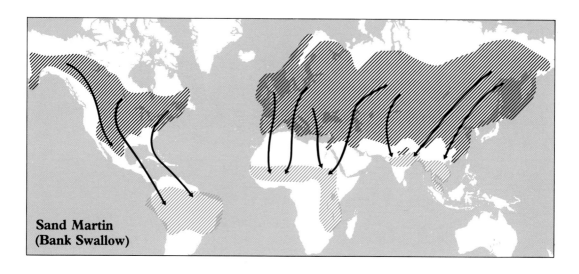

**Sand Martin
(Bank Swallow)**

The Sand Martin

The Sand Martin (known as the Bank Swallow in North America) is a small swallow with brown upperparts and throat and white underparts. It breeds in river-banks, cliffs and many man-made excavations (quarries, pits, railway cuttings, roadside banks etc.) digging its own hole. It breeds right round the northern hemisphere and winters also virtually all round the world (Australia is probably too far for it to reach because of a fairly marked dislike of water crossings!).

At all times of the year it is a very gregarious species and this has encouraged ringers in Britain to concentrate their efforts on Sand Martin ringing. During the 1960's about a third of a million were ringed and the three maps below summarise the information gained on three aspects of Sand Martin migration from the British population.

Left: Colony ringing during the summer combined with autumn ringing at the roosts enabled the pattern of autumn movements in Britain to be mapped in some detail. The bias towards the southeast may be to minimise the length of the mandatory sea crossing.

Middle: Once across the Channel the British Sand Martins move down the Biscay coast of France, through Spain and into North Africa. There are very few ringing records in the Sahara or further south but it seems that these birds winter in Senegal.

Right: The spring return seems to occur in a broad front across the Sahara – probably after the birds have spread eastwards along the southern edge of the Sahara during the winter months.

Autumn

Autumn

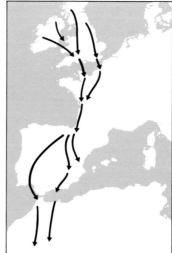

Spring   More eastern route

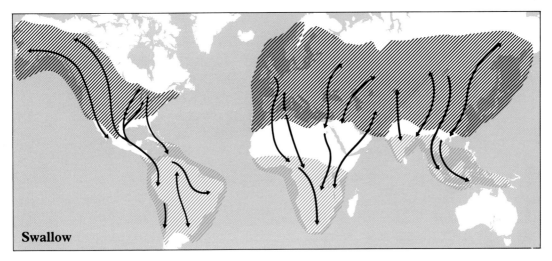

Swallow

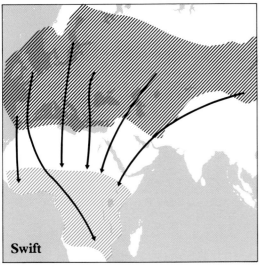

Swift

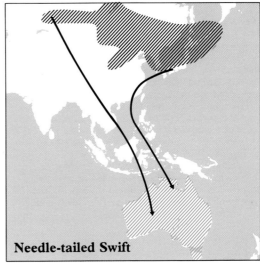

Needle-tailed Swift

Swallow
Perhaps the world's most familiar long-distance migrant, Swallows have both a wider breeding distribution than the Sand Martin and also penetrate further to the south in the winter.

Swift, Needle-tailed and Chimney Swift
These three related species are supremely aerial and yet shun long sea crossings on migration. Details of the Swift's winter distribution, from different parts of its summer range, are not well known but many ringed birds from Britain have been found in the southeast part of their winter range. The Needle-tailed Swift is one of the few landbird species to move into Australia for the winter. In eastern North America the Chimney Swift is a common summer migrant which winters in northern South America and yet almost all fly west-abouts round the Gulf of Mexico.

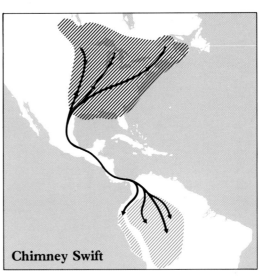

Chimney Swift

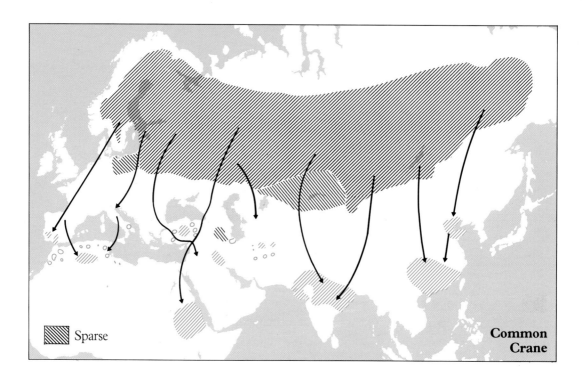

Sparse

Common Crane

The cranes have always been of great significance to man, holding a special place in the culture of such diverse peoples as the Ancient Greeks, the Lapps and the Plains Indians of North America. This is partly because of their great size and superb dancing displays and partly because of the distinctive and unbirdlike cries the migrants make as they pass south in the autumn and north in the spring.

The upper map shows the summer and winter distributions of the most widespread species. Of course the breeding colonies, scattered over remote swamps, are very sparsely distributed and obviously absent from areas with large human populations. The Great White Crane is a very much rarer species that was once much more widely distributed across Siberia. The total world population is now probably rather less than 2000 birds.

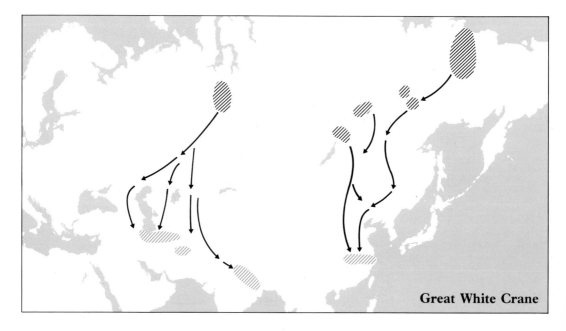

Great White Crane

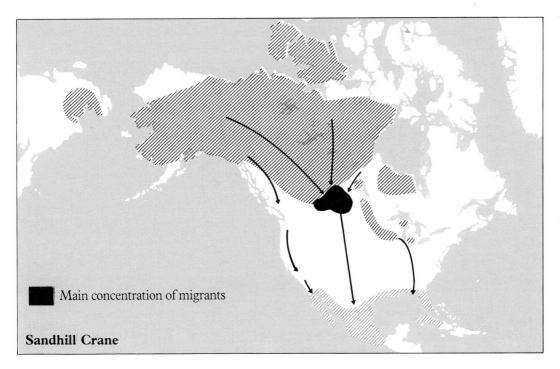

■ Main concentration of migrants

Sandhill Crane

Sandhill Crane

The common crane in North America is the Sandhill with thriving populations in many parts of the breeding range and even some small, sedentary groups to the south. The total population is certainly in hundreds of thousands and considerable damage has often been reported to grain crops exploited by the autumn migrants. As with all the migrant crane species there are particular favoured routes which have been used ever since records were first kept.

Whooping Crane

The gradual build-up in numbers of these magnificent birds has been a conservation success story but there are still less than a hundred alive – a tiny fraction of the flocks that once existed. The map on the left (below) shows the current breeding and wintering areas whilst the other, on the right, shows where they used to be. Each year the wardens in Texas wait anxiously to see how many of the returning pairs have young with them. Most winter on one special reserve in Texas.

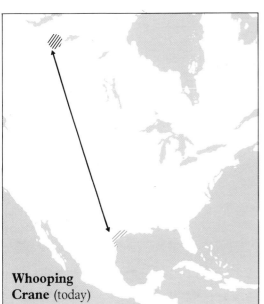

Whooping Crane (today)

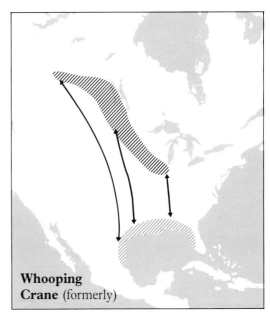

Whooping Crane (formerly)

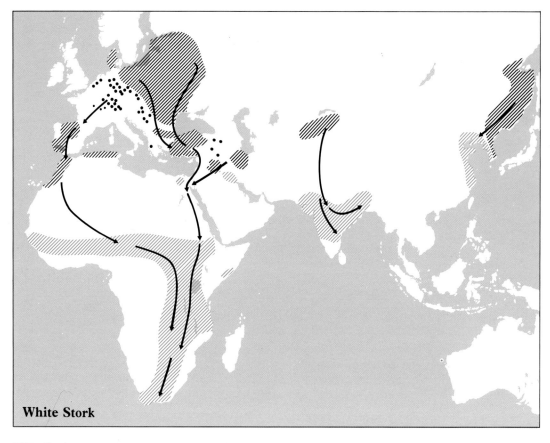

White Stork

White Stork

Most migrant White Storks follow very definite traditional routes along which, in most areas, they are welcomed and pass peacefully. Unfortunately the insidious danger of toxic chemicals has rendered some of these areas unsafe and reinforced a general decline in numbers over much of their distribution in Europe. The soaring birds at places where they concentrate provide one of the most magnificent spectacles in nature.

The annual migrations of the White Stork have been followed in Europe for many centuries. These three photographs show a stork in a typical roof-top nest in a Dutch town (*opposite*); a flock of storks gathering height on the thermals by the Gibraltar Straits (*top*); and standing amidst the dry savannah of central Africa.

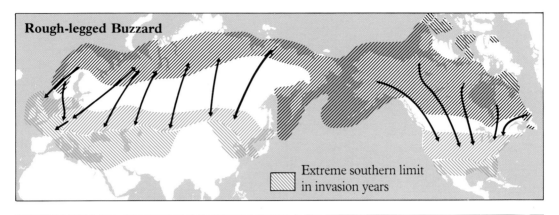

Rough-legged Buzzard

Extreme southern limit
in invasion years

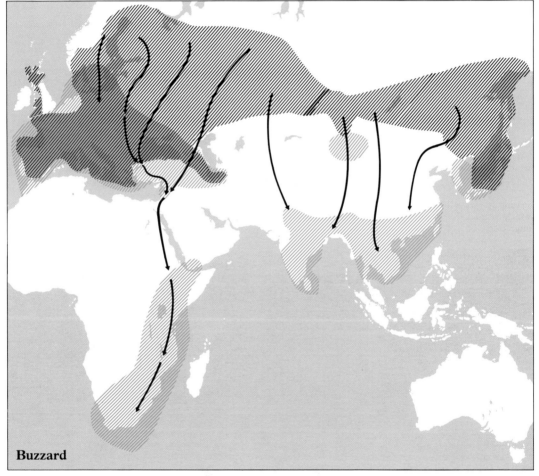

Buzzard

Rough-legged Buzzard
This is a northern breeding species of tundra areas. It
moves southwards every winter but to a varying extent
and only reaches the southern edges of the winter
distribution plotted in years when its prey, small
mammals, are at a low population level.

Buzzard
The western and eastern populations of this woodland
species move much less than those from other areas.
This is an effect of the climate at the extremes of its
range where the influence of the sea warms the winters;
in the continental centre the weather gets far too cold
for Buzzards to stay during the winter.

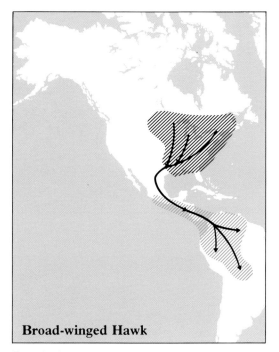

Broad-winged Hawk

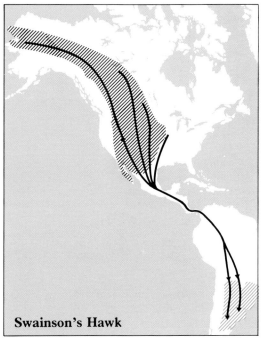

Swainson's Hawk

Broad-winged Hawk
This common woodland hawk streams through the western coastlands of the Gulf of Mexico on fall migration soaring in flocks on the rising thermals or kettles of warming air during the morning. As a soaring species it is dependent on the thin land bridge which connects North and South America to make its migrations.

Swainson's Hawk
This is a bird of prairies and open areas during the breeding season which, like the Broad-winged Hawk, migrates in flocks using the land bridge to make the journey from North to South America. However it continues its migration to reach the temperate parts of eastern South America for the winter.

A Common Buzzard.

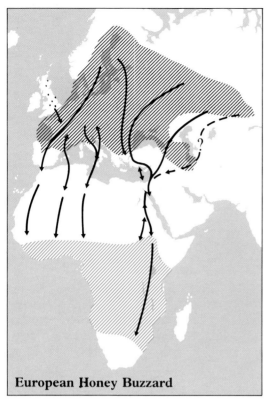

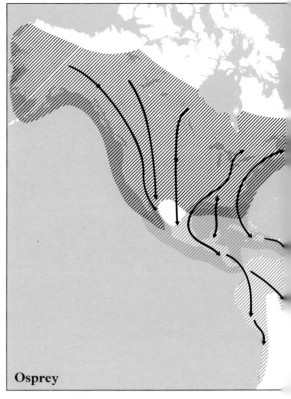

European Honey Buzzard

Osprey

Honey Buzzard

The Honey Buzzard is essentially a European breeding species although two close relatives (sometimes treated as conspecifics) are to be found in Asia. All European Honey Buzzards winter south of the Sahara in Africa. Many of the birds move along the main soaring migrant routes through Gibraltar, the Bosphorus and round the eastern edge of the Caspian Sea. However Honey Buzzards are much more at home with flapping flight than many soaring birds and there are distinct (but generally minor) streams of migrants over other areas – for example through the Balearics and Cap Bon in Tunisia.

Osprey

This magnificent fish-eating raptor is a familiar, if rather rare, bird over most of the world's land not too close to the ice caps. The map shows the breeding and winter distributions of the migratory northern populations and does not attempt to map the sub-tropical and tropical sedentary ones and still less the coastal birds breeding in Australia. It is a much stronger flier than the soaring raptor species so far mapped and rather long sea crossings have no terrors. It therefore ignores the well-defined routes but rather migrates on a broad front. This does not mean that individual birds do not regularly call in at the same places on migration or that particular areas do not seem to be very attractive to migrants Ospreys – both happen and it is much more likely that you will see a migrant Osprey on a reservoir where one was seen last year than one where Ospreys have not been recorded for some years.

Black Kite *(see right)*

This species is found from the Atlantic seaboard of Portugal east to Japan. Over the western half of its range it is, like the Honey Buzzard, a long distance migrant using both the major traditional soaring bird migration routes and the minor ones. Over the eastern half of its range southern populations are sedentary or nomadic and it is only the northern ones which move large distances.

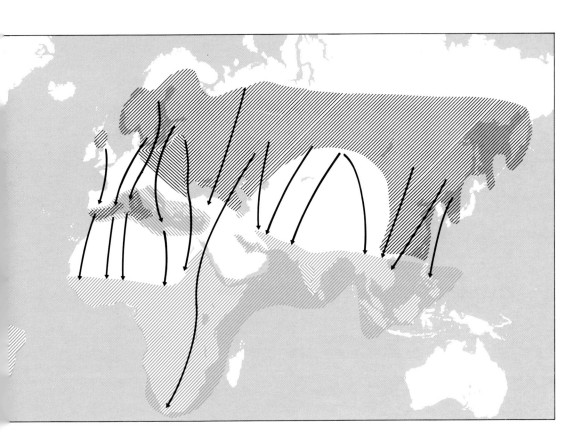

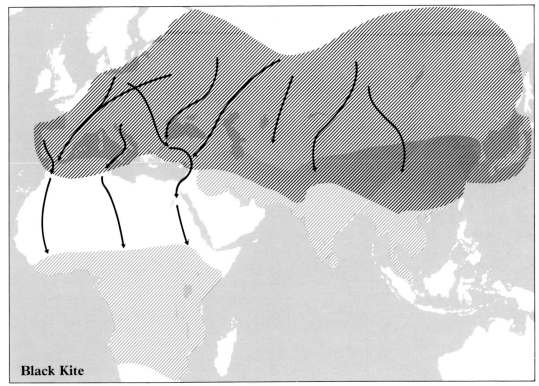

Black Kite

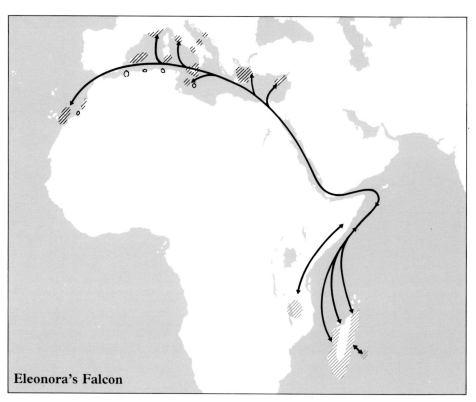

Eleonora's Falcon

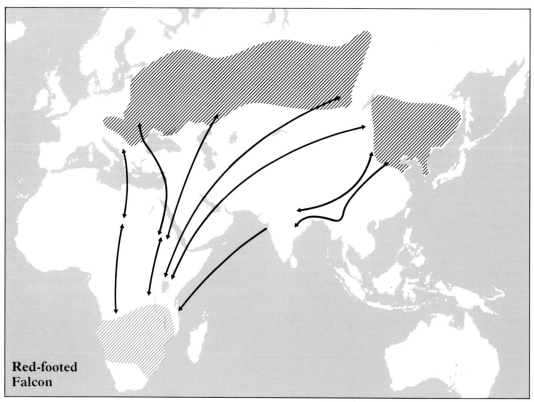

**Red-footed
Falcon**

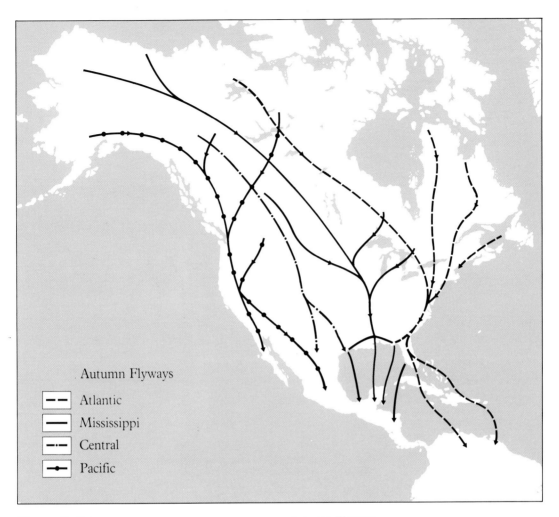

Autumn Flyways

– –	Atlantic
——	Mississippi
–·–	Central
•—•—	Pacific

Eleonora's Falcon

This extraordinary migration route is quite unlike the ones used by soaring birds of prey. The falcons are able to travel long distances in powered flight without having to rely on updraughts to gain height. There is certainly no hint that the western breeding Eleonora's take the direct flight across Africa but all seem to move in the arc round the Mediterranean and down the Red Sea. There are recent indications of a passage across the Horn of Africa and of a small winter population on mainland East Africa.

Red-footed Falcon

The birds from the eastern breeding area are sometimes treated as a separate species (*Falco amurensis*) but is clearly very closely related to the western birds. In any case both winter in the same area of southern Africa. At least some of the eastern birds seem to make a regular very long trans-oceanic migration across the north-western corner of the Indian Ocean.

Autumn Flyways

An enormous amount of research and countless observations of millions of birds have led North American wildfowl biologists to realise that the autumn routes taken by ducks, geese and swans (and some other birds) form a definite pattern. The four flyways are shown on the map which, of course, omits the detail of distribution in the winter and gathering together in the fall. Obviously the species involved, which may winter quite far north, vary at different points along each flyway and the numbers of birds involved – many millions each year in all – vary from area to area and flyway to flyway. Similar migration systems, for the wildfowl, have been proposed for the Old World but many of the physical features of the Old World run east-west rather than north-south as in North America and the pattern, if it exists at all, is much less obvious.

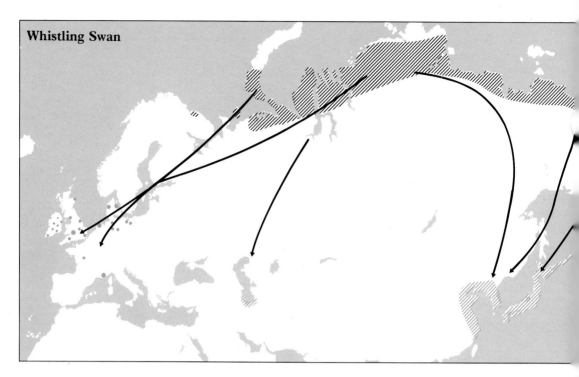

Whistling Swan

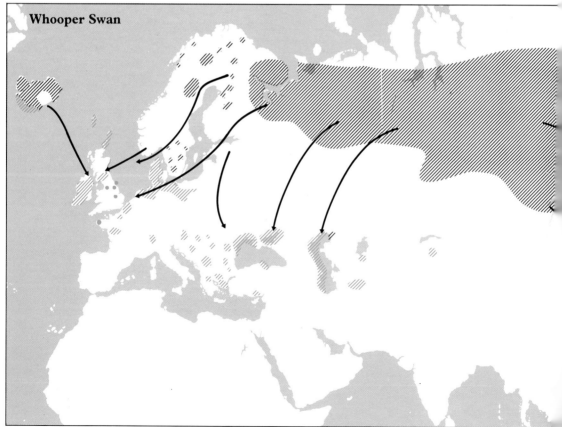

Whooper Swan

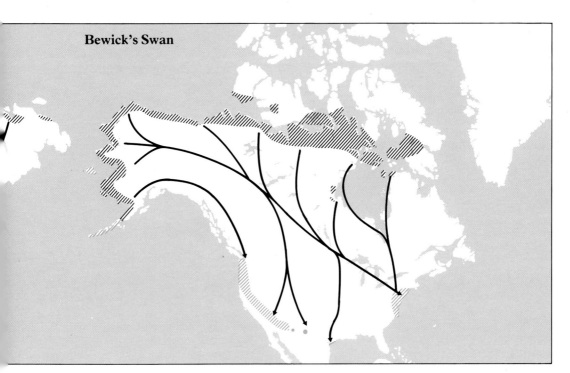

Bewick's Swan

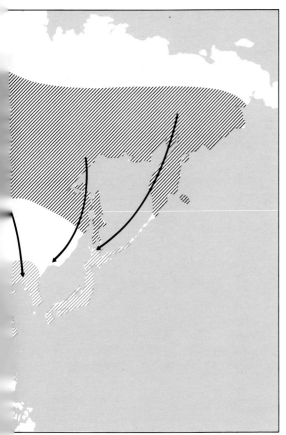

Migrating flocks of swans are amongst the most splendid sights in the bird world – particularly when they are seen moving over a totally inappropriate habitat like a forest or an area of farmland. They are very powerful flyers and do not, in any way, rely on thermals or other updraughts.

The smallest species, Whistling (North America) and Bewick's (Old World) are now considered to be conspecific, breed over a very wide area but in very small numbers, and winter in traditional areas, most of them coastal. This is the smallest species and breeds in the high Arctic where parents and young may set off, in the late summer, for a journey of 5000 or so kilometres. They keep together for the winter and return, together, to the same area in the summer. Studies of individuals, identified from their unique bill-patterns, show that they may arrive at the same winter site in successive winters on almost exactly the same date.

The Whooper Swan, a much larger species and confined to the Old World, breeds further south than the Bewick's but many winter in the same sort of area as them. In most areas they are rather sparsely distributed in the winter in groups of two or three families rather than in the concentrations sometimes found amongst Bewick's where flocks of several hundred are regular.

73

White-fronted Goose

One of the geese with a circum-polar distribution the White-fronted Goose is a familiar migrant in many areas. However since they are very conservative about their wintering areas and migration routes there must be many bird-watchers living within an hour's flight, for the geese, from a traditional wintering area who have never seen them over their home ground. The White-fronts wintering in one area are unlikely to shift to another in a different year but most populations are quite mobile if the weather turns very cold. In addition the spring migration may start early if the weather in the wintering area is mild – if the birds find that there is then a cold snap they may turn back and retrace their steps!

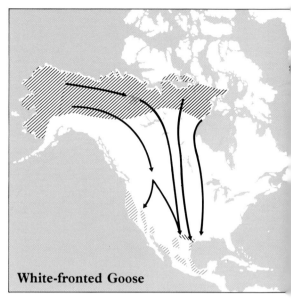

White-fronted Goose

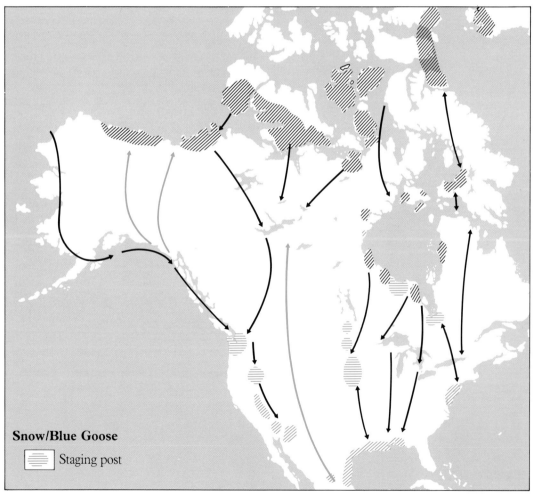

Snow/Blue Goose

Staging post

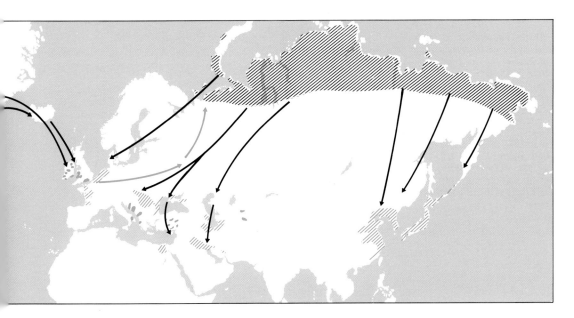

Snow, Blue and Ross's Geese

These geese look like three or more species for the Snow and Blue belong to the same species but vary from pure white birds with black outer wings to dark birds with blue wings (and white undertail). The Ross's is a real species – like a small version of the white Snow Goose – but is very much rarer than the other.

On migration they may occur in vast flocks numbering tens of thousands or more and, at the height of the movement, many thousands may pass the same point in a day in flocks of one or two hundred at a time. The migrating birds often form themselves into a V shaped flock whilst moving and the white undertail is used as a mark point by the following birds. The flocks are also most vocal, particularly at night, as the birds also use sound to keep together. As with the swans the family parties keep together and the same individuals reach the same areas each year.

The long flight south, which may take birds from breeding areas within the arctic circle to winter in Mexico, is too far for the birds to make it in a single trip and so some areas, roughly halfway, are regularly used by transient birds to rest and feed for several weeks. It is these areas, in particular, where really large flocks may build up. In most cases they feed on spilt or spoiled grain but they may do a tremendous amount of damage to individual farmers if they arrive before the harvest has been gathered in.

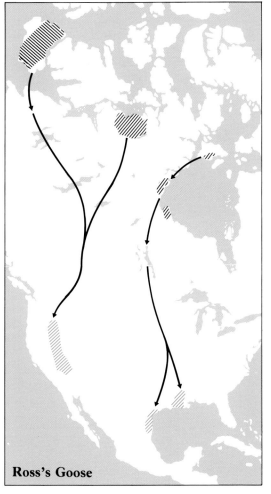

Ross's Goose

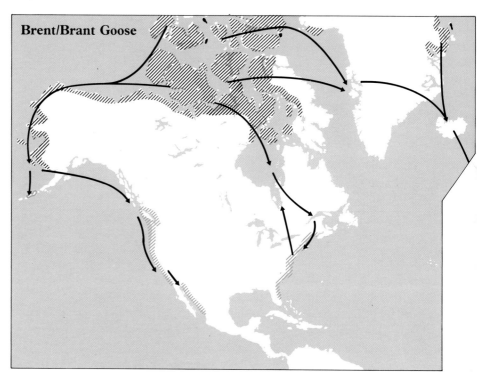

Brent/Brant Goose

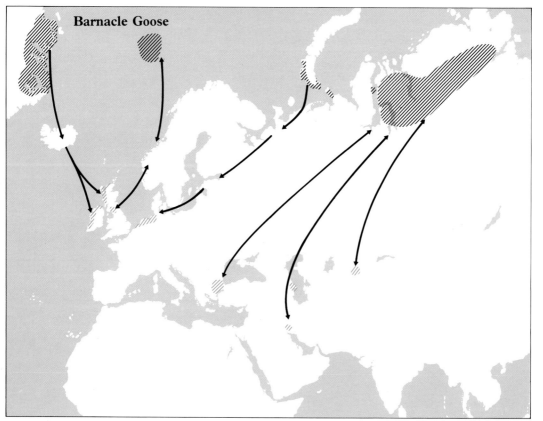

Barnacle Goose

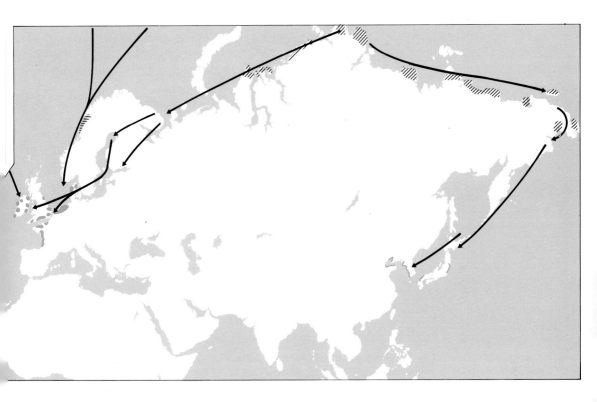

Brent/Brant Geese

These small geese are at home in coastal areas during the winter. Their migrations are quite complicated for they breed in the high arctic and many populations have to move considerable distances east or west to reach the continental shores. Marking of the populations in the Canadian arctic has shown that many birds, some from much further west than Hudson's Bay, move east across the Greenland icecap to Iceland and thence to winter in Ireland. These are the Pale-breasted race of the Brent, those breeding in western Siberia are the Dark-breasted.

Barnacle Goose

There are four distinct breeding areas of this attractive little goose from Greenland in the west to Siberia in the east. Each breeding population has its own winter area or areas with the Spitzbergen population being particularly vulnerable as they all winter on the single estuary (the Solway) on the border between England and Scotland. These birds are within about an hour's flight of wintering flocks from Greenland and yet the two seldom mix and, even then, it is generally only a handful of misplaced stragglers that happen to arrive in the wrong place.

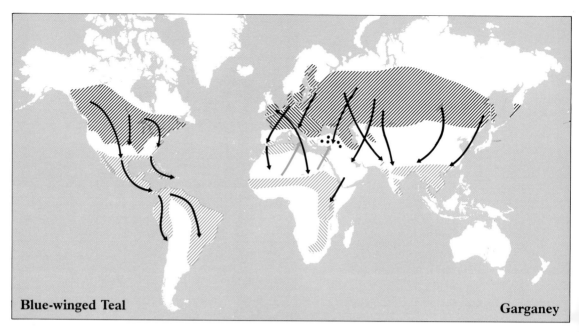

Blue-winged Teal

Garganey

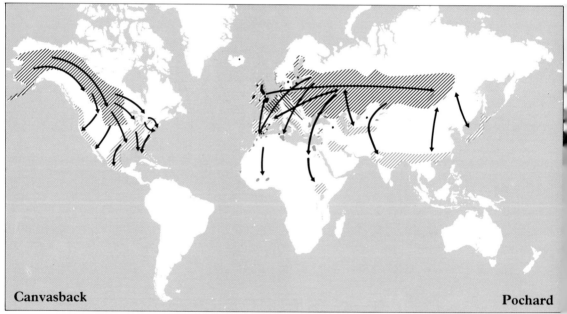

Canvasback

Pochard

Garganey and Blue-winged Teal

These two little ducks look quite similar and are the long-distance migrants, amongst the anatidae, from Eurasia and North America respectively south into and even across the tropics. Over most of their breeding range they are strictly summer migrants although they are very occasionally reported in the winter in the southern part of their range. In common with many duck species they may congregate in large numbers to moult at particularly favourable sites during the autumn.

Canvasback and Pochard

These two diving ducks are forced out of much of their breeding area during the winter when open water freezes. Their movements are southwards and seawards. For the Pochard this is not to the sea itself but to more coastal areas where the warming influence of the sea keeps inland freshwater bodies open. On the other hand the Canvasback, the North American species, during the winter is a duck of coastal bays and estuaries seeming to prefer salt water sites.

78

Ring-necked Duck

Tufted Duck

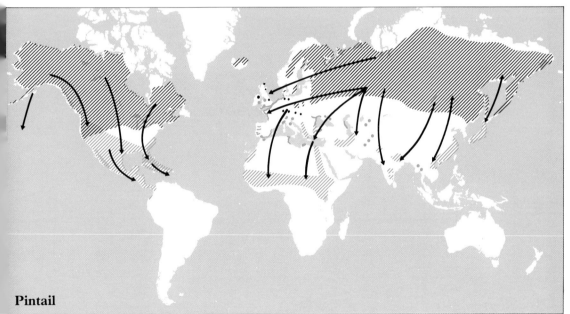

Pintail

Ring-necked and Tufted Duck
These two black and white diving ducks are closely
related but are truly distinct species. Like their red-
headed relatives, the Pochard and Canvasback, the
North American species, the Ring-necked, is more at
home on salt water in the winter than its Old World
cousin. However there are many areas where inshore
bays, with quite salty water, can be found which
regularly harbour flocks of winter Tufties.

Pintail
This species of dabbling duck has a world-wide breeding
distribution and quite a long-distance migration. The
directions are east-west for some birds which move to
coastal areas in the temperate regions but others move
southwards and regularly reach the tropics. It is a
species not frequently found on salt water but makes
trans-oceanic flights and has been recorded on many
islands in the Atlantic and particularly the Pacific.

79

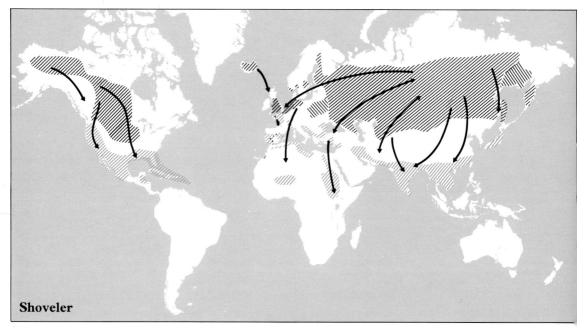

Shoveler

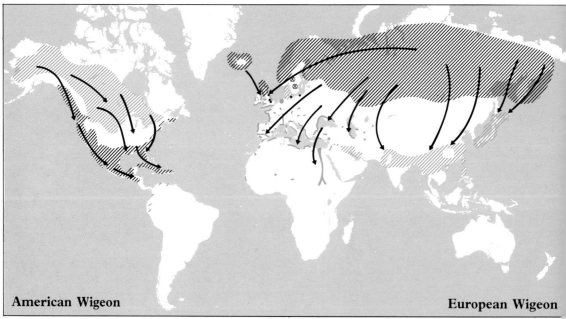

American Wigeon

European Wigeon

Shoveler

This holarctic dabbling duck breeds over wide areas on small waters and along the vegetated margins of larger ones. During the winter many of these areas become frozen and the ducks move out to warmer areas where they may often flock in fairly large numbers. During the spring flocks of wintering birds and transients may be found displaying (and even pairing up) hundreds or even thousands of kilometers from the areas where the displaying birds will eventually settle to nest.

American and European Wigeon

These two dabbling ducks are closely related although the males are rather different from each other. Both species flock in large numbers during the winter and may be found both on fresh water areas and also in coastal bays and inlets. Each species is regularly recorded, in small numbers, along the coastal areas of the 'wrong' side of the Atlantic where they are often found in the company of their local relative.

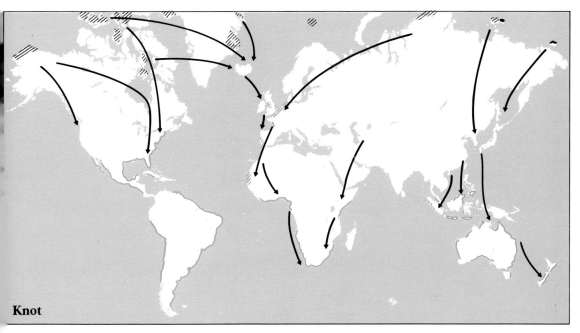

Knot

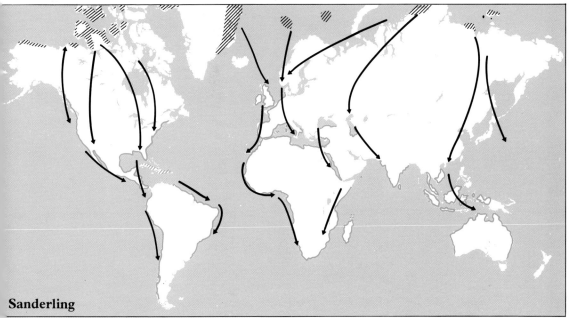

Sanderling

The Knot and Sanderling

These two high arctic breeding species are very long-distance migrants which both also have wintering populations in temperate northern latitudes. They spend the summer so far north that they are only able to remain on the breeding grounds for the minimum time to enable the young to be raised to their full flying ability. When they find severe weather during the summer and the snows do not melt at the normal time many pairs, often most of the population of the area affected, are unable to breed at all.

During the winter both species are coastal. The Knot is found in large flocks mainly on muddy estuaries whilst the Sanderling is found in smaller numbers along beaches of sand or gravel. They are seldom recorded inland but there is very good evidence that there are many areas where they regularly migrate overland – indeed their travels from the high arctic would be very difficult and much longer if they kept to the shoreline. Ringing has shown that individuals may travel at great speed – some Knot from Britain have been found in West Africa within two weeks!

81

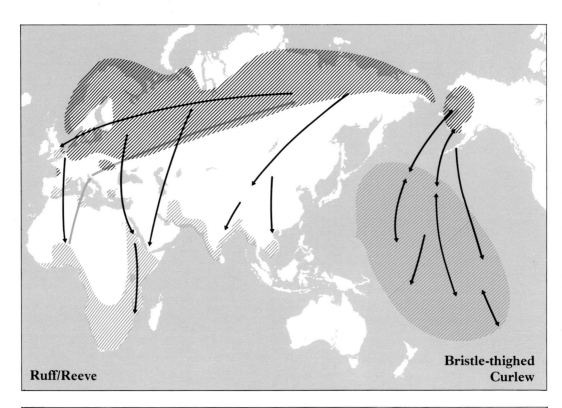

Ruff/Reeve

Bristle-thighed Curlew

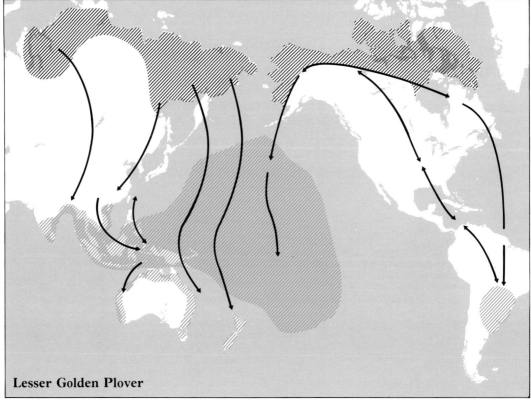

Lesser Golden Plover

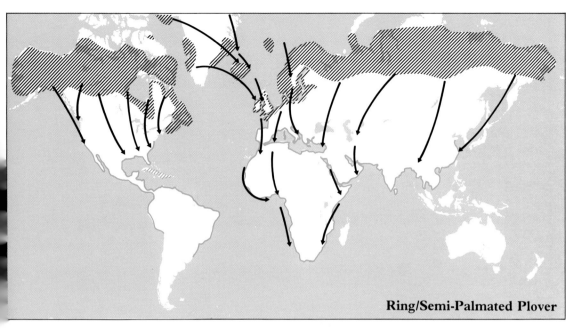

Ring/Semi-Palmated Plover

Ruff and Reeve

This species (the Ruff is the male and the much smaller Reeve the female) is a very long-distance migrant with a long western movement to Britain by some of the eastern breeding population. There is then a movement southwards to the west African Niger inundation zone before the spring movement eastwards again.

Bristle-thighed Curlew

This species breeds in Alaska and winters on the Oceanic islands of the Pacific. Here it has been recorded both feeding along the shore like any respectable wader and feeding on the eggs of colonial terns!

Lesser Golden Plover

This species (different from but closely related to the European Golden Plover) performs some of the longest over-sea movements of any waders. Not only do they regularly winter in Oceania and southward to New Zealand and Australia (probably mostly the Siberian population) but the American birds fly out over the Atlantic to reach their wintering area in South America.

Ringed and Semi-palmated Plover

These two birds (often considered to be conspecific) breed all round the northern hemisphere but winter mainly in coastal sites. As migrants they are regularly seen along rivers and the shingly shores of inland lakes and reservoirs but few remain at such sites for long. Wintering birds reach the southern tip of Africa but Asian birds do not penetrate Indonesia. A few Semi-palmated cross the tropics particularly along the east coast of South America.

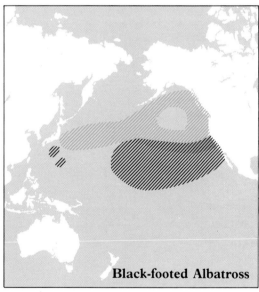

Black-footed Albatross

Black-footed Albatross

This species is now virtually confined, as a breeding bird, to the Hawaiian chain although it formerly bred at several other sites in the north Pacific. The breeding distribution looks very peculiar as it is further south than the non-breeding range – in fact Black-footed Albatross eggs are laid at the end of November so that the chicks leave the colonies in June or July. The two other north Pacific Albatross species are Laysan with a similar distribution to Black-foot but occurring a little further south, and Short-tailed. The Short-tailed is the western Pacific species which is now very rare (colonies extinguished by plumage hunters) but is still sometimes seen off California.

83

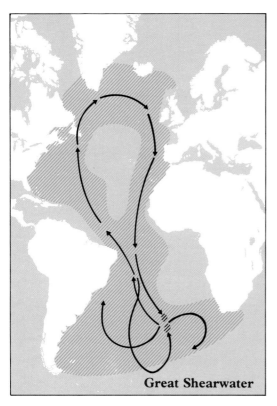

Great Shearwater

Slender-billed Shearwater

Shearwaters and Petrels

These truly pelagic species are often only ever found within sight of land for the small part of their lives when they are breeding. The trans-equatorial movements of the three species of shearwaters involve millions of birds and take them through areas of ocean subject to violent storms. However these birds are adapted to their life-style at sea and are perfectly capable of riding out bad weather. On the left the Great and Slender-billed Shearwaters show quite similar sorts of looping migrations from their breeding area in the southern hemisphere up the Atlantic and Pacific respectively. Above, on the right, is the map for the Manx Shearwater, the common breeding shearwater in Europe, which also has extensive trans-equatorial migration routes. Below on the right are plotted the areas where Northern Fulmars wander. There are few areas in the north Pacific and North Atlantic where Fulmars are not seen at all times of the year for the young birds probably do not come back to land for several years after they fledge. In fact, the Fulmar is a species whose adults probably stay fairly close to the colonies as they may be seen gliding in front of the breeding cliffs for several months in the winter and spring before the eggs are laid. Individual Fulmars must travel incredible distances, compared with the other species illustrated, for, although they do not venture so far from their natal areas they are constantly on the move.

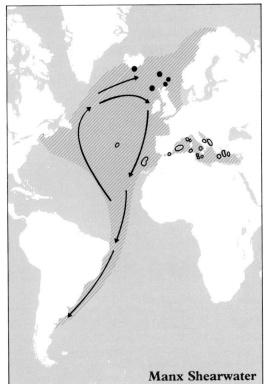

Manx Shearwater

A group of Manx Shearwaters
resting at sea.

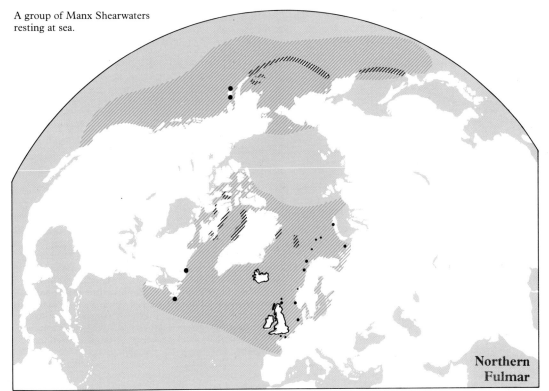

**Northern
Fulmar**

85

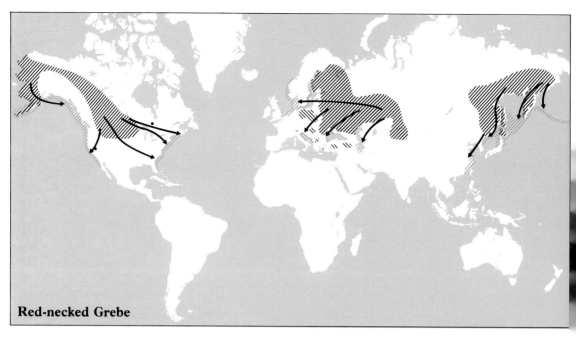

Red-necked Grebe

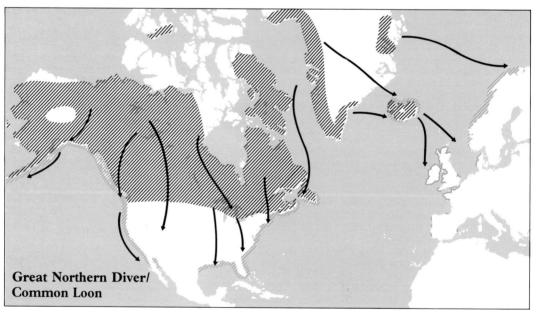

Great Northern Diver/
Common Loon

Red-necked Grebe
This large grebe has a rather patchy distribution but, for
a bird which seems to be an inexpert flyer, it undertakes
long movements to winter in coastal waters. Such
movements are forced on it for the breeding areas
freeze completely in the winter. Migrant birds may
regularly be seen on intervening fresh water areas and,
during the winter, storm-driven birds are sometimes
found inland.

Great Northern Diver or Common Loon
This bird, with the hauntingly beautiful call on its remote
arctic breeding grounds, is also forced out during the
winter by freezing conditions. Most winter in coastal or
estuarine waters but a few stay on southern rivers and
lakes in the United States. Although some reach
Europe after long flights across the sea the diver's food
is generally found in fairly shallow waters and so they
are not found wintering far out to sea.

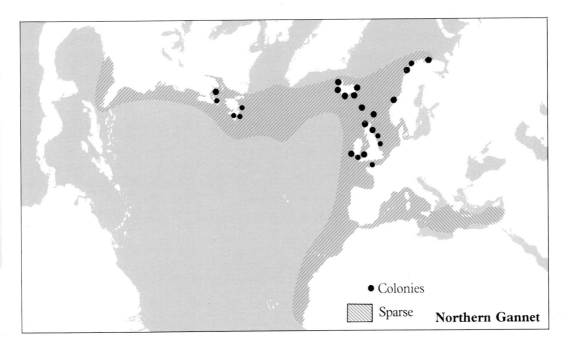

● Colonies

░ Sparse

Northern Gannet

Northern Gannet

Northern Gannets are a familiar sight all round the North Atlantic but, as with the grebes and divers, they are not generally found in mid-ocean. There are great differences in the movement patterns between young, immature and adult birds (the youngsters move much further and so most records from the southern areas are of the brown youngsters). These movements, for a hypothetical individual, are plotted on maps on page 157.

The Storm Petrels

Storm petrels are tiny birds which appear to be much too puny to survive bad weather at sea. On the contrary they are much happier away from land and are seldom seen from the shore – even near their breeding colonies which they only visit at night to escape the attention of predators. There are many species world-wide whose distribution is generally known only from the collection of specimens (like the Least off western Central America) but the European Storm Petrel movements are a little better known as several British and Irish ringed birds have been found in South African waters.

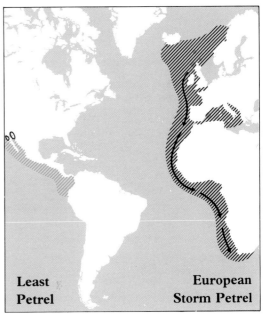

Least Petrel

European Storm Petrel

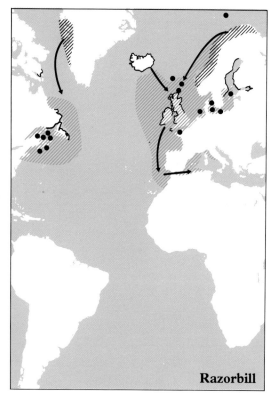

Razorbill

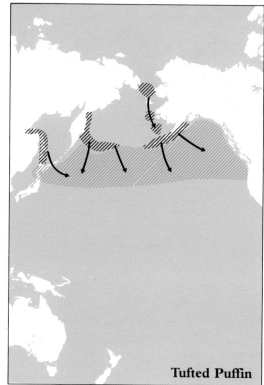

Tufted Puffin

Razorbill and Tufted Puffin

These two auk species may move quite a long way southwards for the winter to reach warmer waters, at least
from their northern breeding areas for both are still
present in the vicinity of their southern colonies through
the winter. The Razorbill is predominantly a coastal
species but the Tufted Puffin feeds and lives in mid-
ocean for much of the winter.

Bonapartes and Lesser Black-backed Gull

The patterns of movement shown by these two gulls are
similar to many other species which breed inland but
winter mainly in coastal areas. For the gulls the Great
Lakes and such huge, but land-locked, waters as the
Mediterranean, the Caspian and the Black Sea also
provide good winter habitat. The Lesser Black-backed
Gull moves exceptionally far, for a gull, and they are
regularly recorded in the Rift Valley of Africa astride
the equator.

Kittiwake

This is the Black-legged Kittiwake, the only species in
the Atlantic but one of two in the Pacific. Although
structurally a gull it is a cliff-ledge breeding species
which congregates in huge colonies even in the
northernmost areas. During the winter they spread over
vast areas of the North Atlantic and North Pacific and
take up a truly pelagic way of life. Ringing records of
British birds show a regular passage across the Atlantic
and, within their overall range, individual populations
may indulge in very complicated regular movements.

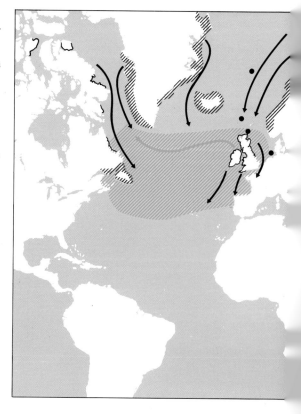

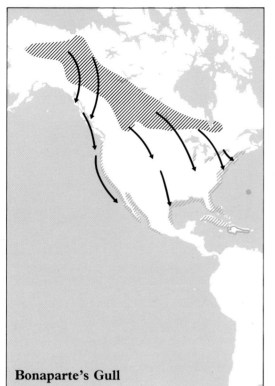

Bonaparte's Gull

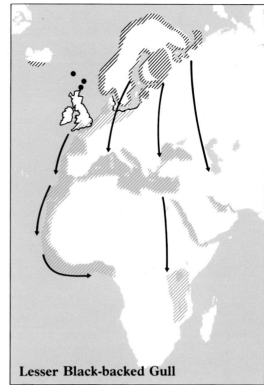

Lesser Black-backed Gull

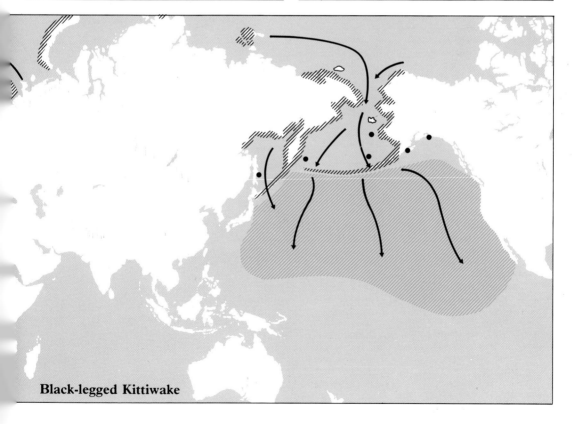

Black-legged Kittiwake

An Arctic Tern.

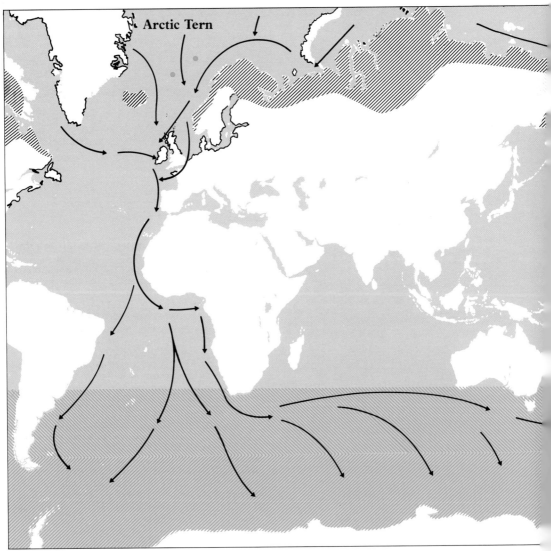

Arctic Tern

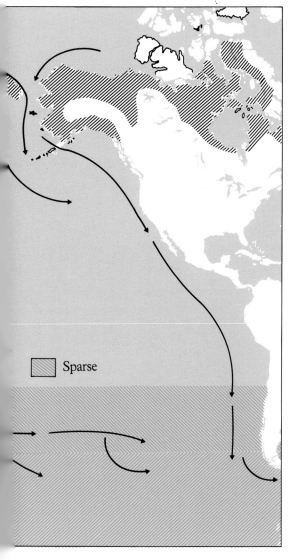

An Arctic Tern in flight.

The Arctic Tern
This is, by general consent, *the* migrant. Arctic Terns, which breed north of the Arctic circle, migrate southwards to reach the Antarctic pack ice and so enjoy more daylight in a year than any other creature. The map only shows the routes known or inferred from ringing recoveries and it is probable that, over the years, a lot more fine detail will be added. Some British ringed nestlings have been found, in spring, in the region of the Urals and their presence here has been taken to indicate a mistake on their journey northwards. Most were fairly young birds and may have 'missed' the Atlantic but migrated northwards up the Indian Ocean. The urge to travel north then overcame their aversion to moving over land – which, in any case is probably not very marked – and they carried on northwards.

This is one of the few migrations which has partly been investigated by computer simulation. The results of the analysis, which was done by feeding the computer with known facts about weather and winds and reasonable assumptions about the tern's flight, are shown on page 203. They agree closely with the ringing records and with observations made on Arctic Terns on migration. Not only is this interesting in the context of the tern's flight but it also gives an insight into the method whereby migrations have evolved and traditional routes become established.

Sparse

Arctic Skua

In many respects the patterns of Arctic Skua migration mirror those of the Arctic Tern. This is because both species breed in the same arctic areas and the Skua is regularly a pirate of the tern at sea. In fact Arctic Skuas are very adept at stealing food from many species of terns and so do not only move where the Arctic Terns go. In the Atlantic the main tern route down the west coast of Africa is certainly an important route but the eastern seaboard of America (used by terns other than Arctic) is also used. In the Pacific large areas of the ocean have wintering and passage Skuas since there are vast colonies of local terns for them to harry: the same is partially true of the Indian Ocean.

In fact although their main feeding method is piracy from terns and other birds the Arctic Skua is also capable of scavenging from the sea surface or from stranded debris. It can also catch its own living food at sea – provided it is very close to the surface. As with many seabird species, since most have a long pre-breeding period, many Arctic Skua immatures will remain in southern areas through the northern summer although most, even in their first summer, make some movement northwards.

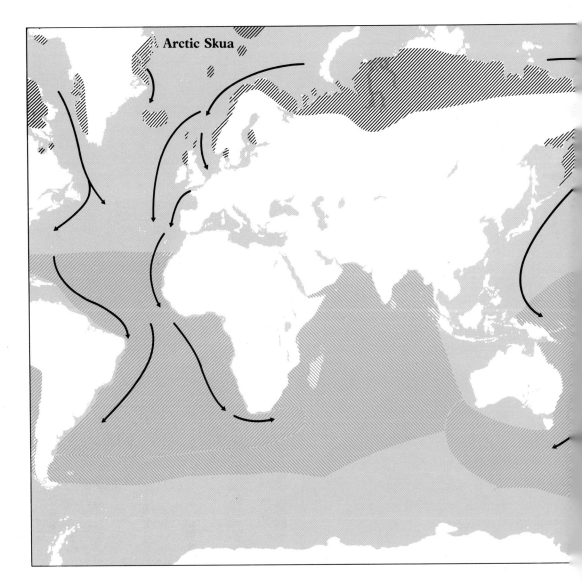

Skuas, such as this Arctic Skua, are experts on
'pirating' food from other seabirds. Photograph right
shows a Skua chasing a gull in an attempt to force it to
relinquish its catch. The photograph below shows an
Arctic Skua on its nest.

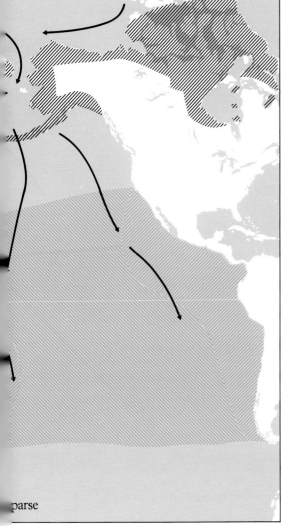

parse

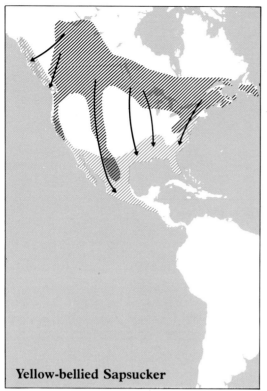

Yellow-bellied Sapsucker

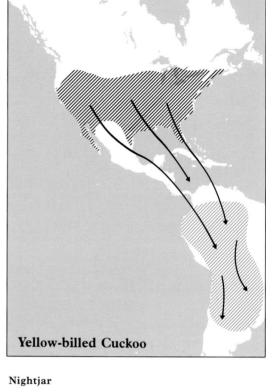

Yellow-billed Cuckoo

Yellow-bellied Sapsucker
This attractive little woodpecker has developed a specialist feeding technique. It drills parallel rows of small holes in the bark of living trees so that the sap flows from them. Later they return to feed on the sap and on insects attracted to it. During the winter the trees in their breeding area are dormant and the sap would not run. They are therefore forced to move out and spend the winter much further south and penetrate as far as the west coast of Panama.

Yellow-billed Cuckoo
Both the common cuckoos of North America, this species and the Black-billed found only in the east, migrate south to winter in South America. They are very similar species but the Yellow-billed has rufous in the wing and more white in the undertail. This is not one of the parasitic cuckoos but rather makes its own, rather flimsy, nest and lays a clutch of three or four eggs and looks after them perfectly normally. It is a fairly late migrant and a successful vagrant to Britain and Europe with a bird or birds recorded in late October or early November every few years.

Nightjar
The 'European' Nightjar actually also breeds far across Asia and some migrate long distances to winter in the far south of Africa. All are trans-Saharan migrants and winter in areas where there are many local nightjars but also, of course, many of the large insects that they hawk at dusk and during the night.

Turtle Dove
This attractive small dove is the only long-distance migrant in its family to breed in Europe. Almost all are full trans-Saharan migrants although some breed well to the east in Asia. There is no indication of more distant movements and all the Turtle Doves congregate for the winter in the sub-Saharan savannah region.

Wryneck
This primitive woodpecker is the lone member of its family in Europe to undertake long-distance migrations. It has a superbly marked cryptic plumage and lacks the other woodpeckers' stiff tail feathers. Its feeding habits are less specialised and it is happy to feed on the ground or in bushes. In Britain it is one of the migrant species seen every autumn on the east coast having crossed the North Sea. Few now breed in Britain for its populations in England have almost become extinct over the last 50 years.

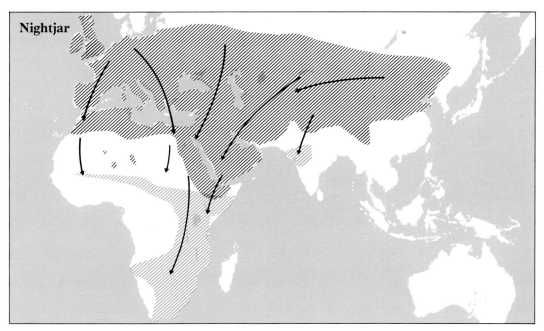

Nightjar

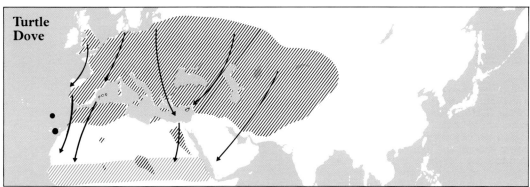

Turtle
Dove

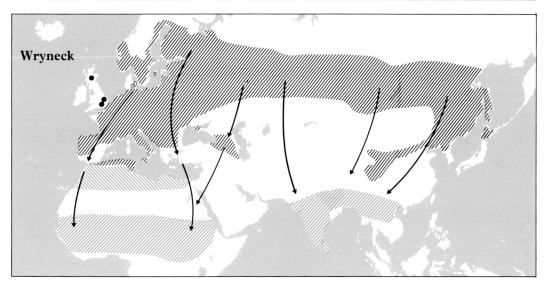

Wryneck

95

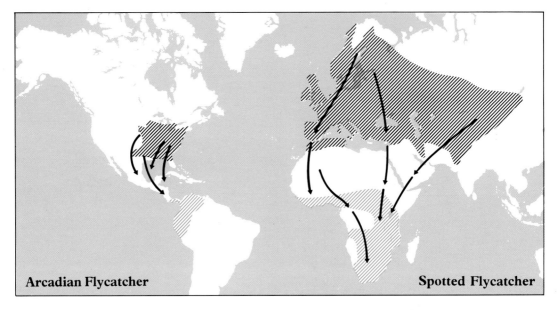

Arcadian Flycatcher

Spotted Flycatcher

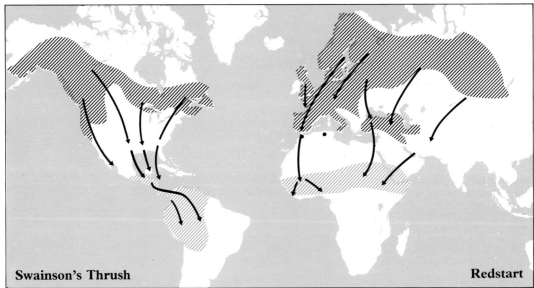

Swainson's Thrush

Redstart

Arcadian and Spotted Flycatchers

Although these two species are not closely related their
flycatching activities would not be very profitable during
the winter in the areas where they breed. They are
therefore both migrants to spend their winters in much
warmer areas where their insect food is plentiful.
During the summer the Arcadian lives in damp,
deciduous woodland and is one of a group of very
similar species (the genus *Empidonax*). The Spotted
Flycatcher is found in all sorts of habitats through
Europe – forests, woodland, hedges and gardens –
where there are perches from which it can make its
feeding flights. All the *Empidonax* species of North
America winter in Mexico or further south.

Swainson's Thrush and Redstart

These two species are both small thrushes. The Swainson's
is one of five similar species found in the woodland
under-storey through much of North America. They
are mainly insectivorous although most also eat berries,
particularly in the autumn. Of the five species only the
Hermit Thrush regularly winters in any numbers in
North America, but only in the southern states. The
Redstart, a brightly coloured chat completely unrelated
to the American Redstart which is a woodwarbler, is an
insectivorous woodland bird which winters in sub-
Saharan Africa. It is a familiar migrant, particularly in
the autumn throughout southern Europe and North
Africa.

96

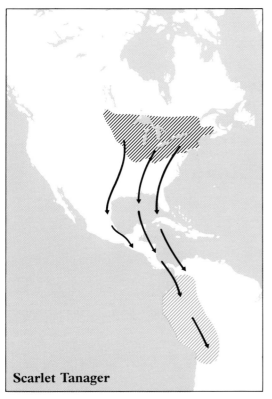

Scarlet Tanager

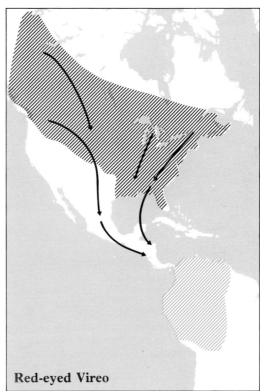

Red-eyed Vireo

Red-eyed Vireo

This is a very common species in deciduous woodland. The vireos are rather like the warblers but have a rather thicker bill and are generally not so acrobatic and perky. Vireos mostly feed by picking insects off foliage and seldom, if ever, indulge in flycatching. All North American species of vireos (apart from Hutton's on the west coast) are migrants although some Solitary and White-eyed Vireos remain in the southern states for the winter. Once more the vireo family is wholly American and there are no exact equivalents in the Old World.

Scarlet Tanager

This is the north-easternmost representative of the larger tanager family – mainly a tropical group of wholly American birds. In the west its place is taken by the Western Tanager which breeds well into Canada. Both birds migrate far to the south for the winter – the Western not necessarily as far as the Scarlet. They are woodland species, the Scarlet being found in both broad-leaved and coniferous forests but the Western is mainly a bird of spruce, fir and pine. They have no real equivalent in the Old World.

A Spotted Flycatcher showing its typical upright posture.

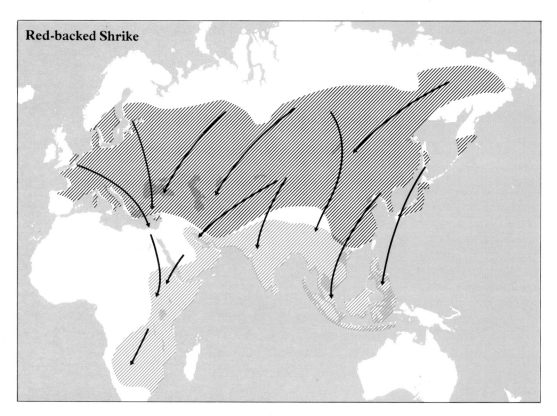

Red-backed Shrike

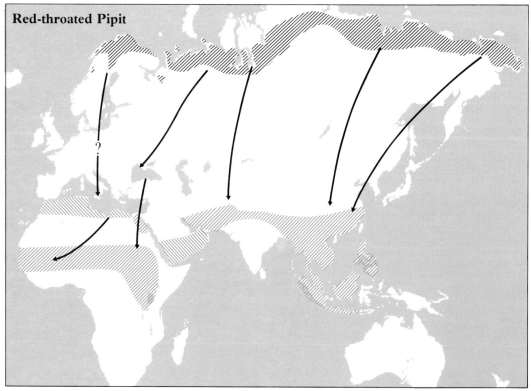

Red-throated Pipit

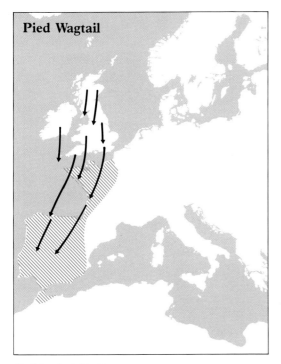

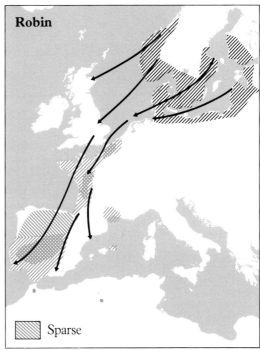

Sparse

Red-backed Shrike
These striking birds, although passerines, are carnivorous and feed on small mammals and birds as well as large insects. They are migrants southwards from their breeding grounds which extend all the way across Eurasia. In the west their populations have recently slumped and they are one of the few species to migrate southeast across the eastern end of the Mediterranean from western Europe.

Red-throated Pipit
The pipits are small ground-feeding birds mostly of open areas. The Red-throated breeds in the tundra areas across northern Europe and Asia. It is the longest-distance migrant of all the pipits and most in the west cross the Sahara whilst many in the east reach Indonesia.

Pied Wagtail
This map, based on the ringing recoveries of Pied Wagtails marked in Britain and Ireland, shows the movements of that breeding population. In fact the Pied Wagtails breeding in Britain are a little different from continental birds (known as White Wagtails). The species as a whole breeds over all of Europe and Asia even with breeding records in Alaska and on Greenland. It is only by using ringing records that the movements of a single population within the whole can be charted.

Robin
As with the Pied Wagtail map beside it the Robin map charts movements plotted from ringing recoveries. Once again the breeding range covers vast tracts of Europe and spreads into Asia. The movements shown are partly of the small part of the British breeding population which moves away for the winter but mostly of passage migrants from Fenno-Scandia which are caught and ringed on autumn passage. The winter records of these birds shows their main concentration to be in Iberia. The summer recoveries are mostly in Norway, Sweden and Finland. Other recoveries from these countries show passage also through France, Italy and other parts of Europe to North Africa. The Robin is, of course, a small insectivorous bird about a fifth the weight of the American Robin (which is about the size of a European Blackbird).

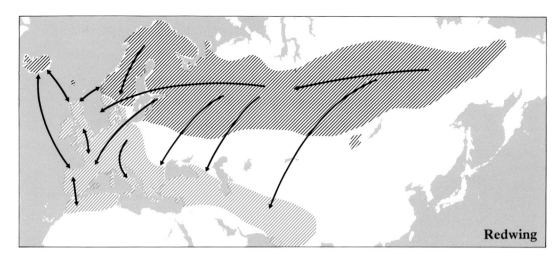

Redwing

Brambling

Redwing

This attractive, medium-sized thrush is a winter visitor
to western and central Europe from its breeding
grounds in the Northern forests. Ringing records show
a very complicated series of movements and that birds
coming far to the west in one year may travel south or
even east in subsequent years. They are very liable to
move further in the winter if their food supply of
berries runs out or the weather becomes much colder.

Brambling

Bramblings are finches which congregate in the winter to
feed in flocks on seeds. They are particularly attracted
to the irregularly available nuts from Beech trees and,
if there is a very good crop in any area, their numbers
may build up into millions. They are closely related to
the Chaffinch but show a distinctive white rump in all
plumages.

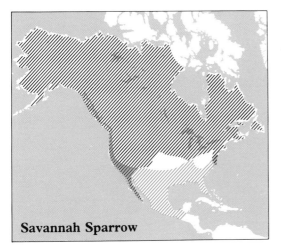

Savannah Sparrow

This drab-streaked sparrow of open grassland is forced to move southwards for the winter by the covering of its possible feeding areas by the winter snowfall. A few populations, in the west, are probably sedentary.

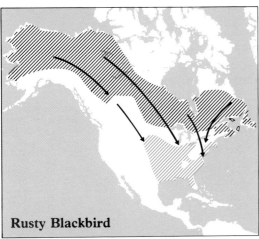

Rusty Blackbird

The Rusty Blackbird is an icterid, a member of the American group which includes the Meadowlark, grackles and cowbirds. It is a bird of swampy woods rather than a flocking species of open areas like the other blackbirds. Even the western breeding birds, found through much of Alaska, move far to the east for the winter.

Bramblings foraging for food on a woodland floor.

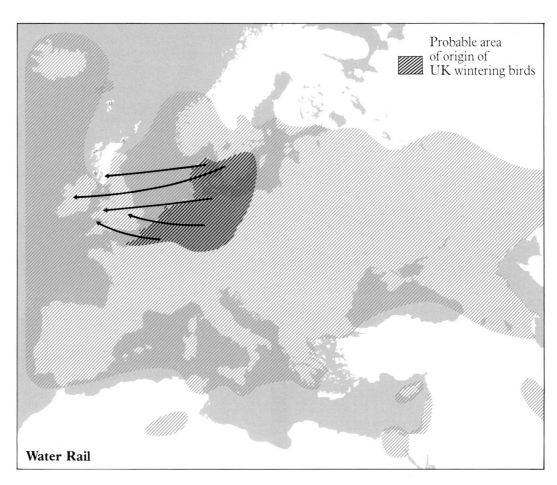

Probable area
of origin of
UK wintering birds

Water Rail

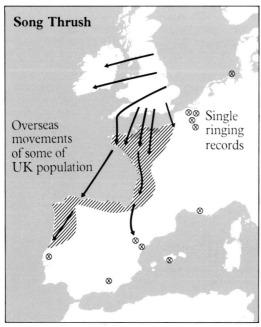

Song Thrush

Overseas
movements
of some of
UK population

Single
ringing
records

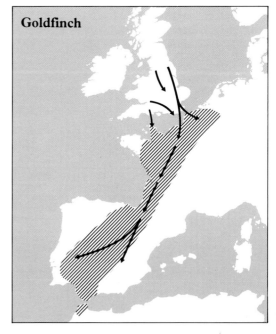

Goldfinch

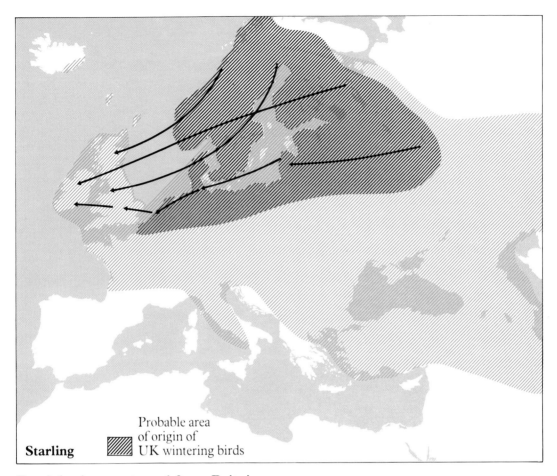

Probable area
of origin of
Starling UK wintering birds

Partial migrants to and from Britain

These four species are all familiar birds over most of
Britain and Ireland throughout the year. Ringing
records have made it clear that all four are, nonetheless,
involved in large-scale movements involving Britain.
This sort of movement affects many species through the
world but is obviously impossible to depict, over the
whole of a species' range, without resorting to large
numbers of individual maps.

Water Rail

This is a very shy bird of reed-beds and swampy areas of
rank vegetation over all but the northernmost part of
Europe. In Britain it is rather seldom seen although its
presence is signalled by its unearthly cries. Ringing
records have shown that Water Rails from as far east as
Sweden and Poland come to Britain for the winter.

Song Thrush

This map does not show the passage through Britain of
the few continental birds that cross the North Sea.
Rather it shows the movements of British birds out of
their native haunts during the winter. Even in mild
winters a few birds move but, if the weather is
particularly cold, large numbers may move not only
westwards to Ireland but southwest to Devon and
Cornwall and southwards to France and even Spain.

Goldfinch

This is a very attractive finch with red on the head and
gold in the wings. It is different from the American
Goldfinch but quite a close relative. In Britain flocks
may be found on thistles and teasels during the winter
but many native birds leave in the autumn and spend
their winters in France and Spain. They return quite
late in the spring when flocks of newly arrived birds
are often to be found along the south coast. The
regularity of their migration and its destination was not
discovered until ringing started.

Starling

This familiar bird, now introduced to and present over
most of North America, is actually a very strong
migrant over much of its Old World range. The
British population is a bit nomadic but certainly not
migratory but, during the winter, Starlings from much
of northern Europe stream over the North Sea to
winter in huge flocks over the whole of Britain. Other
populations go to France, Iberia and other parts of
southern Europe. There are none left over most of
northern continental Europe during the winter. Indeed
their return in the spring is welcomed as an indication
that winter is over.

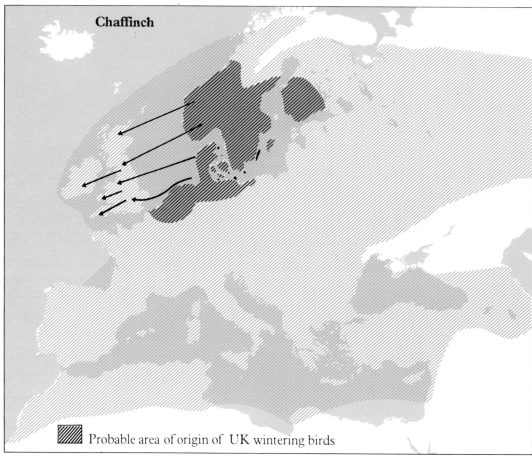

Chaffinch

▨ Probable area of origin of UK wintering birds

Winter movements to Britain

The Chaffinch and Goldcrest are two more species with extensive British and Irish breeding populations that are also winter visitors in large numbers. This had long been realised for both can be found in large numbers on the western edge of the North Sea in the autumn. The Chaffinches breeding in Britain are very sedentary and ringing returns have been received of birds more than ten years old which were still within two hundred metres of their original ringing place. They may be found in woods, hedges and gardens breeding wherever there are tall enough trees to act as song posts. The Goldcrest, a close relative of the North American kinglets, used to be thought of as a bird of coniferous woods but is present in deciduous forests too.

Both species move into Britain and Ireland from Denmark, Norway, Sweden and Finland. Ringing recoveries of both also come from Belgium, the Netherlands and northern Germany but these may be birds bred further north passing through on passage. Goldcrests from the southern Baltic may be breeding there or, possibly, on passage. This pattern of movement will, of course, be repeated over the rest of their range in northern Europe. Indeed, for many species, the birds which travel across the North Sea from Scandinavia to Britain are actually undertaking a shorter migration than those from further east.

Moult Migration

Most migrations are annual movements from breeding areas which become unsuitably cold in the winter to warmer parts of the world. There are some special migrations, particularly among the waterfowl, which are to special moulting grounds where the birds congregate to undertake their complete autumn moult. Ducks, geese and swans, unlike most other species, become flightless as all their main wing and tail feathers are lost at the same time. The moulting grounds therefore have to be safe from predators, good feeding areas and sheltered – for the birds are not able to move away.

The Shelduck is a large, goose-like duck which breeds along the coasts of much of Europe and, in some areas, at inland sites. It nests in burrows in sand or shingle and produces large broods of young which are often found, joined together, in creches of several dozen chicks. A few adults stay with the youngsters but almost all the adult birds move away to moult in the Helgoland Bight. There is also a small moult area in Bridgewater Bay in Somerset.

The Canada Goose, only introduced to Britain about 200 years ago, has increased greatly in numbers over the last twenty years or so. Within that time a regular moult migration of the non-breeding adult birds to the Beauly Firth has developed from the main concentration of nesting birds in the north Midlands.

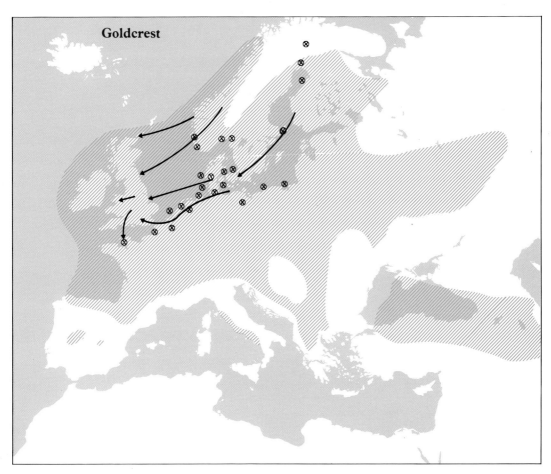

Goldcrest

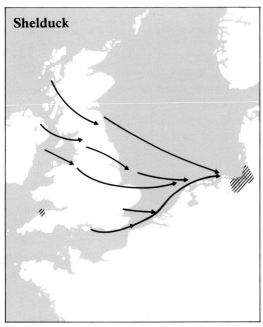

Shelduck

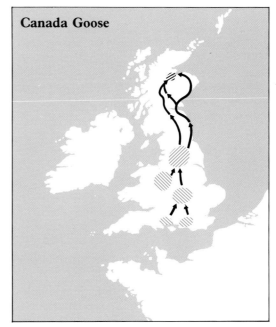

Canada Goose

Making the journey

The simple statement that a bird migrates from one place to another implies a capability, on the part of the bird, to undertake two related but separate tasks: it has to find the way and it has to be able physically to make the flight. Finding the way, or navigation, has always been seen to be a major source of interest and has had much theoretical attention from generations of scientists (see the next chapter), but actually being able to fly the distances they migrate is a phenomenal achievement for many long-distance migrants.

Just as a car or aeroplane needs fuel in the form of petrol, diesel or kerosene the bird must have fuel to be able to fly. The bird's source of energy is, of course, the food that it eats but, to be able to fly long distances quickly the bird cannot take in food regularly during its flight. It must be able to store the necessary energy before it migrates, just like charging up a battery or filling a petrol tank. Much research has been done on migrating birds by scientists in recent years and their results have confirmed what the traditional bird-catching cultures already knew – migrating birds are exceedingly fat. The fat, laid down subcutaneously (under the skin) and within the body-cavity, is the source of

power and the long-distance migrant has to prepare itself for its major flight by laying down this fat over a period of weeks. For the processes by which this is done and the metabolic implications of them see the chapter on Annual Accounts.

For the moment let us take a 'fully fuelled', fat migrant warbler and consider the problems it has to face in completing its migratory flight. First and foremost it must be able to fly. Flapping flight is a very complicated manoeuvre but is learnt by a young warbler within two or three weeks of leaving its egg. As we shall see, the problems it faces are compounded by the extra load of fat which it has taken onboard as fuel and by the different constraints imposed on it by its migration. The small bird, in its bush, feeding on more or less agile insects needs to be able to make short, sharp forays with the premium on acceleration and speed. The same bird, as a migrant, will need to maximize its flight potential by consuming as little fuel as possible per unit of distance covered. Secondly, it must know when to set off, not only in the seasonal sense – i.e. to make an autumn migration – but also by taking advantage of weather conditions that are likely to favour it or, at the very least, are unlikely to be directly opposed to its completing its migration. Thirdly, once it is airborne, apart from knowing where it should be going it needs to be able to take advantage of any helpful wind component which may be blowing at one height and not another. Finally, when its particular flight stage is finished, it should be able to recognize where to make a landfall in conditions which are favourable. This is a simple thing if the flight is short and the ecology of its reception and departure area are similar but very difficult if the flight has taken the bird to an ecologically different part of the world.

Flight

The power of flight has fascinated man for thousands of years. The myth of Icarus flying too close to the sun with wings made of feathers and wax that melted in the heat illustrates this fascination. Man has flown, with mechanical aids of varying complexity, only for the last century or so. Even the most recent flights of *Gossame Albatross*, the incredibly light fixed-wing monoplane powered by human muscle alone, must be classed as freaks compared with the accomplished efforts of the birds. Indeed any attempt by any living creature trying to provide lift for flight by flapping with the types of muscles which have so far evolved is

doomed to fail if that creature weighs more than about 20 kilograms. Gliding flight, using either a structure of membranes stretched between bones and feathers or, for humans, a hang-glider made of plastic sheeting and metal spars, is possible for much larger creatures. The scaling factors involved are easy to understand: for every two-fold increase in length there is a four-fold increase in area, the factor which supplies most lift, but an eight-fold increase in volume, the factor which controls weight. For muscles the major factor which controls power is the number of fibres proportional to the area of cross-section of the muscle. Thus, as a muscle becomes more powerful it will quadruple in weight as the power is doubled. Chickens, ducks, geese and turkeys are all quite big birds and they contain a lot of good meat. Those succulent cuts from the breast are the massive muscles needed to allow the birds to fly using flapping of the wings rather than gliding. Gliding flight is rather different. Instead of providing the lift to keep aloft through the strenuous flapping of the wings all that is needed is for them to be kept braced – generally requiring a twentieth or less of the power needed for flapping. The bending and flexing of the wing (or tail or even feet) for control purposes in gliding flight also uses much less power than the very considerable

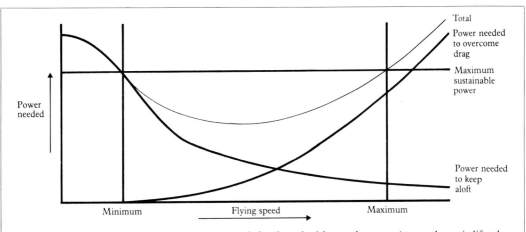

As a bird's speed of flight increases the power needed to keep it airborne decreases (as aerodynamic lift takes over) but the power loss to drag increases. Only where the combined power curve is less than bird's maximum possible power output can flapping flight take place. For hummingbirds, minimum speed is zero since they are able to hover.

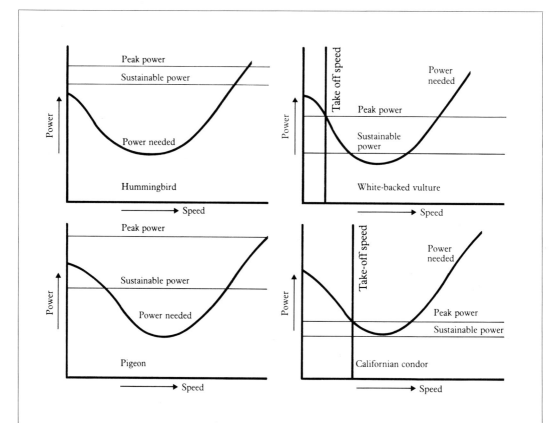

The power equations for these four species show how much harder flapping flight becomes for bigger birds. The hummingbird can easily fly and even hover, a pigeon can only take off using more power than it is able to sustain, the other two cannot take off unless there is a wind or something high to jump from and the Californian Condor cannot even undertake sustained flapping flight.

control needed on every stroke of the wing, up and down, in flapping flight.

Let us consider powered flight, that is with flapping wings, in more detail. It is generally agreed that there are four main components to the power requirement of the flying bird:

1) To provide lift to keep the bird aloft.
2) To overcome the drag (friction) of the bird both as it moves forward and as the wings flap.
3) To overcome the inertia of the wings – accelerating them at the start of a stroke and stopping them at the end of it.
4) To power the bird's metabolism – both ordinary life functions and the power to circulate the blood and ventilate the lungs.

The lift is partly generated through the strokes of the wings. These may seem, at first sight, to be helping to support the bird only on the downstroke but analyses of the trajectory taken by the wingtip has shown that even the simplest flapping flight is not an up and down motion but generally a complicated figure of eight, with the upwards part of the stroke also generating lift. The bird's wings are, of course, constructed so that they can readily be 'feathered' on the upstroke – that is their resistance to the air is lessened. However, this is not the only way that the bird obtains lift. The passage of the wings and the bird's body through the air generates aerodynamic lift because of the cross-section of the bird as it passes through the air. The flow of air over the bird and over an aircraft's wing are obviously very different but there are similarities for all to see. Indeed the most famous British fighter of the Second World War, the Spitfire, owes the basic shape of its wings to the designer watching Herring Gulls flying. The basic principle of the aerofoil is that the air pressure above the wing is reduced – sucking the wing upwards – whilst the air pressure below the wing is increased – pushing it upwards. The exact flow of the air over the wing is very important, for any turbulence will reduce the wing's efficiency. Of course, unlike the fixed wing of an aircraft, the bird possesses a wing whose contours can be altered at will. Thus, where the aircraft designer has to stick with a rigid wing design, it is quite possible for the bird to alter its wing to suit the task it has to accomplish. The analogy between a bird wing and an aircraft wing can be taken a stage further: the control surfaces in front of the main aerofoil, the slots on an aircraft, which help to give control at low flying speeds, are present in the bird's wing as well. This is the function of the small 'bastard wing' (or alula) on the front of the bird's wrist.

The drag of the air over the bird is an important factor affecting the efficiency of flight. The design of cars has started to take this into account and models are tested in wind-tunnels to make sure that air-flow over the design is smooth. The bird's feathers make a very efficient surface which may control turbulence by ensuring that the flow of air very close to the bird is as smooth as possible. Any turbulent flow will decrease the efficiency of the bird as a flying machine. However, there are, with any bird, circumstances when the efficiency of flight, with regard to the amount of lift, speed or range, is a positive handicap. For instance, for a Kestrel to seek out its prey, it needs to be able to hover in mid-air. Another example is when a bird coming in to land needs as slow a speed as possible for a safe touch-down. In these circumstances turbulent flow may be a positive advantage in diminishing forward speed. However, stalling speed, that is the speed through the air at which lift can no longer counteract the weight of the bird, is the important quantity here. For most species, in the absence of headwinds, which allow for speeded up flow of the air over the bird whilst it is stationary in respect of the ground, really low speeds are only possible with very great effort – at take-off and possibly at landing (some come in to land quite fast!). A series of experiments with pigeons, that gave them mild but unpleasant shocks through their feet, showed that they were physically unable to take-off more than half a dozen times in quick succession; the effort was too much for the relatively small amount of energy available to the muscles used. It is rather like forcing an athlete to sprint 100 metres several times in quick succession.

The two power requirements we have so far looked at are dependent on the size and shape of the bird – particularly how much it weighs. Much more important than this is the dependence on the speed of flight. The faster the bird flies the better its efficiency as a flying object – the power it has to expend in flapping to gain lift decreases as the aerodynamic lift induced by its speed increases. Unfortunately the amount of drag which it experiences also increase as it speeds up and, since it increases faster, there is, for any bird, a theoretical upper and lower limit to its flying speed using flapping flight. In fact a further drag effect has been omitted (see graph on page 107) – the resistance of the air to the flapping of the bird's wings. This drag is quite important and so is the power required to overcome the inertia of the wings. Both are difficult to estimate but will grow with increased flapping speed and, between birds, size of wing. Indeed one may readily test this by waving a fan up and down first slowly and then faster; it becomes much harder to do as the speed of flapping increases. The final power component, the energy needed to keep the internal functions going, is relatively stable for a wide range of flight speeds for any individual species.

Now the power requirements are known, we need to consider just what the migrant bird must do to be able to fly as far as possible with its fuel reserves fixed. There are only a limited number of options available to it once it starts: it can fly at a fast or slow speed, with a consequent effect on its power consumption; it can fly at a high or low altitude, so experiencing air at different densities and temperatures (also with varying amounts of oxygen available); and, finally, it can alter either or both of these variables to take advantage of tail-winds or to minimize head-winds. These decisions are taken by the bird in a perfectly natural way. However, the decisions taken will alter through the course of a long flight as the bird's mass steadily decreases as fuel is burnt. One effect of this is to alter the cruising height so that, as fuel is burnt, the best height, for maximum range, increases gradually. Observations of migrants flying at night have suggested that this does indeed happen but it is 'not proven' at the moment. In principle the partial pressure of oxygen at a high altitude is the factor which restricts the 'best' height for birds to fly; theoretically, some long-distance migrants, half-way through their flight, might do best as high as 8000 metres. There are, however, very few records of small passerine migrants at heights like this.

The theoretical considerations which govern the best speed for a bird to fly to conserve its 'fuel' are quite simple. If it needs more power to fly faster to overcome drag whilst, at the other extreme, it also needs more power to fly slowly (it is not getting the benefit of aerodynamic lift) then there must

be a speed in between which will enable it to maximize its potential (see graphs on page 108). The general shape of the power curve so far shown seems to indicate a simple solution: the maximum range will come at the bottom of the power curve. However, this is not the case as the very slight increase in power needed to fly faster is compensated for by the extra length of journey achieved. The best possible speed for cruising is given by the tangent from the power curve which passes through the origin (see the graph on this page). The graph shows the situation in still air, but if there is a wind blowing, as is generally the case, the graph needs to be altered to take into account the real distance over the ground accomplished by the bird. This can be done by shifting the origin of the graph to the left if the bird is being helped by a tail-wind and to the right if it is being hindered by a head-wind. The resultant tangents, giving the *best speed* for the *maximum range*, show that it flies slower with a tail-wind and faster against a headwind. This is exactly what one would imagine would be the case but the amount by which the best speed is varied turns out to be much less than the wind speed it is encountering. It is also obvious that the effect of head- and tail-winds will be very much greater on the slower flying species than those capable of faster speeds and thus, in general, have higher maximum range speeds.

In the calculations of 'maximum range speeds' very complex equations are used to make allowances for the vast range of different factors involved. These are still being developed and, whilst it is generally agreed that it is along the right lines, the exact solutions are not known. For example, the flight of most small migrants, when they are not migrating, is very different from the characteristic migration flight – generally known as 'bounding flight'. The birds take a series of quick, continuous flaps and then fold their wings for about the same length of time. However, with larger birds (bigger than a Starling) the folded wing part of this cycle is taken up by gliding and takes up a much longer part of the time. These two characteristic means of flying during migration

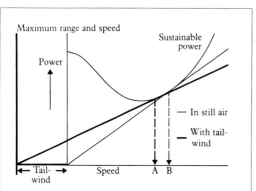

With a tail-wind a flying bird can go further by flying slightly slower (and so staying up longer) than in still air.
A = Optimum speed for maximum range in still air.
B = Optimum speed for maximum range with tail-wind.

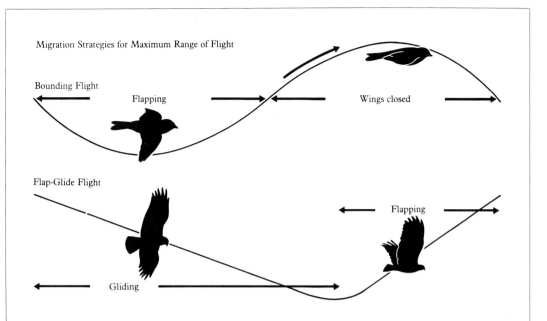

Migration Strategies for Maximum Range of Flight

Bounding Flight

Flapping

Wings closed

Flap-Glide Flight

Flapping

Gliding

When migrating small birds undertake a 'bounding flight' strategy (upper diagram) with their wings closed for half the cycle. This is because their energy losses to drag outweigh the aerodynamic lift they would gain by having their wings out. Larger species make a lot of use of gliding between bouts of flapping. (Vertical and horizontal dimensions of the diagrams are not to scale.)

actually give rise to rather different rates of energy expenditure: waders, for example, seem to be able to do rather better than the small warblers. In even larger birds flapping flight may almost be dispensed with on migration in favour of soaring flight. As we will see later this is a very different strategy from powered flight. It has been suggested that these special migration flight strategies violate some of the principles of the equations used for predicting maximum range speeds but the few direct measurements which have been taken seem to support the theoretical results. To the layman the closure of the wings during the non-powered phase of 'bounding flight' seems to be a disadvantage for the small birds; the equations show that, for birds of Starling size and less, the extra drag induced by the open wings more than negates the extra lift generated by their aerofoil profile. The following table gives maximum range speeds calculated for some species at their normal weights – i.e. without an appreciable fat reserve.

These figures would have to be modified if the bird was carrying a heavy load of migratory fat or other extra weight. Colin Pennycuick, who has been responsible for the development of many of the fundamental equations, worked in Africa. He produced calculated power curves for a White-backed Vulture with its crop empty and the same bird carrying a load of 1140 grams of food. The maximum power it can produce is theoretically too little for it to be able to maintain flapping flight with the extra weight; this was indeed the case and the unfortunate bird, after disgorging its 1140 grams of dead Zebra, was able to fly off under its own power. However, the figures are not critical for small birds, which are regularly able to double their weight to carry 50 per cent or even more fat at the start of migration. Clearly it is not possible for such large species as geese and swans to do this as they would not be able to fly at all. This means there is a very definite restriction on the potential flight range that such large birds have for a single migratory flight. There is some evidence that they are particularly aerodynamically efficient, having

	Species	Speed (metres per second)	
House Martin	*Delichon urbica*	8 m/s	18 mph
Swift	*Apus apus*	9 m/s	20 mph
Common Tern	*Sterna hirundo*	9.5 m/s	21 mph
Red Kite	*Milvus milvus*	10.5 m/s	23.5 mph
Starling	*Sturnus vulgaris*	12 m/s	27 mph
Kittiwake	*Rissa tridactyla*	13 m/s	29 mph
Woodcock	*Scolopax rusticola*	14.5 m/s	32.5 mph
Homing Pigeon	*Columba livia*	16 m/s	36 mph
Cormorant	*Phalacrocorax carbo*	17 m/s	38 mph
White Stork	*Ciconia ciconia*	18.5 m/s	41.5 mph
Mallard	*Anas platyrhynchos*	19 m/s	42.5 mph
Razorbill	*Alca torda*	19.5 m/s	43.5 mph
Greylag Goose	*Anser anser*	22 m/s	49 mph
Bewick's Swan	*Cygnus columbianus*	24 m/s	54 mph
Whooper Swan	*Cygnus cygnus*	26 m/s	58 mph

maximized their shape for this purpose. The calculations for a Whooper Swan on a normal migration leg from the Outer Hebrides in Scotland to Iceland, a distance of 720 kilometres, show that to do the trip on a fat load of 10 per cent it must be very much more efficient as a flying machine than a pigeon. For smaller birds the flight range potential, when they are at a load factor of 50 per cent fat – a figure achieved by a wide variety of passerine species – is often in the region of 2000 kilometres in a single hop. For instance a Ruby-throated Hummingbird may have a still-air range of 2300 kilometres but is greatly affected by adverse winds – even as little as 5 metres per second would reduce its range to 1000 kilometres and double this would stop it from making any headway at all.

Implied in the equations are some factors which the birds will almost certainly take into account. These include, as we have seen, the change in maximum range speed as fuel is used up which, incidentally, may be as much as 40 per cent higher at the beginning than at the end. It also means that power consumption, and thus fuel consumption per unit of distance covered, is much less at the end of a flight than the beginning. For a small migrant starting with 50 per cent extra fat, a gram of fat will take the bird 2.5 times the distance at the end of its flight than it would at the beginning. We have also seen that the greatest

efficiency may be reached at high altitudes; if the birds are to reach these heights they will certainly gain height gradually rather than spiral up steeply. The work used in gaining height like this is recovered by the bird as it loses height whilst gaining distance at the end of its flight. Another strategy to maximize their endurance, if they are lost over inhospitable terrain or caught up by strong winds, is to 'throttle back'; they may be able to remain airborne, whilst flying very slowly, for a much longer time than they would at their maximum range speed. It has been suggested that the American small passerine migrants which regularly reach Northern Europe may have adopted this strategy during cross-Atlantic storms.

So far there has been little mention of the means by which soaring migrants are able to accomplish their flights. The soaring bird, generally much bigger than the small fat-depositing birds, is able to travel long distances expending very little energy, provided that the correct conditions of lift are available. With land-birds like Cranes, birds of prey, storks etc. this is generally through following traditional pathways, taking advantage of 'standing waves' of air on hills and mountain ranges and by minimizing sea-crossings. In Europe the main routes are channelled east and west round the Mediterranean and in America they follow the

A White-backed Vulture thermalling over Mount
Kilimanjaro in Kenya.

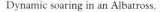
wind direction

Dynamic soaring in an Albatross.

This Wheatear has taken an opportunity to rest during its long migration and was photographed at the open porthole of a cabin on the RRS *John Biscoe*, 200 miles west of Ushant, France.

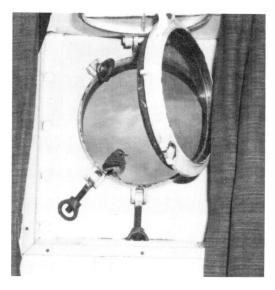

isthmus round the western edge of the Gulf of Mexico. In most instances these migrations take place in the morning and early afternoon after the sun has warmed up the ground to produce thermals (ascending currents of air). The technique is very simple and is used by long-distance sailplane pilots. An area of rising air is found and used to gain height. Then the slow descent rate, using the efficiently designed wings for gliding flight, is used to make as much distance as possible, in the correct direction, *towards another area of rising air*. This technique is often known as 'thermaling' and the birds circling within a thermal are collectively known as a 'kettle' in America. Even if the thermals have no clouds topping them, following birds will often be able to identify the rising air simply by seeing the other birds already using them. The standing waves created on rising ground under certain weather conditions can be used in just the same way and they are responsible for much of the channeling effect of such mountain ranges as have significant hawk migration sites. The best strategy for the birds is, of course, to find a range running in the direction that they wish to travel with a wind at right angles to it. They are then able to ride the updraught and peel off sideways to the next one.

Soaring is not, however, confined to the land-birds. Vast numbers of seabirds use a very special technique called 'dynamic soaring' to put the energy of the wind blowing over the waves of the sea to use. The absolute masters of this technique are the albatrosses but many of the larger tube-noses, the shear-waters, fulmars and petrels, also use it. Basically the bird uses the wind current to climb into the wind. It then makes a high leeward turn and gains ground by gliding with the wind whilst losing height. After making a low turn in the trough of a wave it starts the cycle again. Although it sounds complicated to see the bird actually doing it at sea is to realize that nothing could be simpler, at least for the bird. Many of the species that use dynamic soaring methods, if the wind is really strong, are able to use the rising air behind a ship to be sucked along. Many of them will tag along with a vessel which is apparently going in the right direction for hours or even days on end.

Weather

The consideration of flight techniques has, to a certain extent, taken account of some weather conditions. The effect of head- and tail-winds on the maximum range cruising speed is clear and it is likely to be taken into account by the migrating bird, provided it is able to detect head- or tail-wind effects. Weather will also have an over-riding effect on soaring birds: cold wet conditions when they should be migrating will leave them sitting around in trees, sometimes in large numbers, waiting for the weather to warm up and thermals to form. At sea, bad weather for a 'dynamic soarer' is no wind at all. By

tacking they are almost always able to make some headway if there is some wind but totally calm conditions will often see birds, which would otherwise be migrating, sitting on the sea in flocks waiting for the wind.

For small migrants, with a relatively low maximum range cruising speed, it is very important that they should not try to battle against headwinds but should take advantage of tail-winds. This means that quite apart from sophisticated navigation systems and astonishing feats of physical endurance the migrant has also to be a weather forecaster. In fact there are some relatively simple rules which help small birds to make the correct decisions about the weather prospects when they set out. In eastern North America the fall is generally heralded by a series of lows which regularly arrive from the west. Birds which wish to set out with a tail-wind only needs to identify within which sector of the weather system they are to be able to make the correct choice. In Britain fine anti-cyclonic weather is generally the best for migrants to start their migration. However, the classic conditions for observing large numbers of birds at the coastal bird observatories occur when anti-cyclones over Scandinavia coincide with a frontal system taking cloud across the North Sea. This means that the birds which set out with clear skies to make their migration somewhat west of south to Iberia tend to be brought west across the North Sea by easterly winds and then are lost in the thick cloud and fog. Most of these birds are young and it is possible that experienced adults do not make such mistakes often. These conditions were thought, before the start of radar observations, to induce the birds to fly downwind until they reached land. Such 'down-wind drift' may take place in some circumstances but it is now thought that the majority of the birds coming to shore in such conditions are simply lost.

We are almost always concerned only with the weather as it affects us on the ground, not realizing just how different conditions are above our heads. For one thing it is very much colder. Migrant birds are regularly flying in temperatures as low as $-15°C$ over North America in the autumn. These conditions may actually help the birds which are working hard and have to get rid of surplus heat as they fly. Indeed it has been calculated that most species would be unable to undertake sustained flapping flight in warm countries as they would overheat. However, at all latitudes, the air at altitudes of two to three thousand metres above ground level is always below $10°C$ and so birds are able to carry on simply by ensuring that they are flying high enough – or at night. It is probable that the early morning flights of diurnal flapping migrants, like finches and buntings, are to ensure that they do not overheat. A further consequence of the low temperatures is that they are able to prevent dehydration. Their metabolism means that fat stored as fuel, when used, creates water. This they need to excrete but, for it to happen, they must also breathe in quantities of oxygen. The calculations show that the volume of air that they must inhale to provide sufficient oxygen would, at warm temperatures, take out with it, as it was exhaled, large amounts of moisture from the lungs. At temperatures of about $10°C$ or less they are able to reduce this water loss to less than the amounts produced by metabolic processes. Some physiologists believed, at one time, that the flight potential of migrants might be restricted by their loss of water through respiration.

The influence of weather systems on migration is crucial but most species are well adapted to cope. Indeed, even those that have an apparently poor record – being prone to 'wrecks' of birds in front of storms or during foggy weather – are probably only losing a small proportion of the population (generally young birds). Such 'rush' conditions are much more frequent on the eastern seaboard of the United States of America with birds which were over the ocean coming to land. Here too the majority are probably able to cope and carry on with their migration. In any case, whether they are migratory or sedentary, most small birds in the northern hemisphere have rather low survival rates. It may typically only need a single adult and a single youngster from each pair of adults and their five young who set off in the autumn to return in the spring for populations to be maintained.

Altitude of flight

The physiological advantages of flying at different altitudes have already been discussed. However, the differing directions of the wind at different heights can be of great importance to migrants. None, as far as we know, are able regularly to reach the height at which the jetstreams start to blow. These very fast winds, which can reach speeds of 500 kilometres per hour (300 mph), may start as low as 9000 metres. They are not easily predictable from ground-level but are used to great advantage by the airlines. It is often worth diverting flights over quite long distances to take advantage of such winds. In the northern hemisphere they flow west to east in a sinuous band round the earth between the latitudes of about $40°$ and $60°$. Also, winds at a lower level than these jetstreams may be much stronger than at ground-level and often blow in rather different directions. Research with radar and other methods of recording the height at which the migrants are flying has shown that there are often concentrations of birds at the heights with the most favourable wind components.

This observation, which has been confirmed by some but not all research teams, has some very interesting implications. The birds as they fly, imbedded in a body of air moving in a particular direction, do not have any local frame of reference with which to gauge the direction of the wind. In order to sense what is happening they must be able to perceive a direction, relate the heading on which they are flying to it and then realize that they are being drifted by the wind. Various theories have been suggested as to how they do this; the most likely mechanism simply involves the birds seeing a fixed feature on the ground and then using their navigational sense to keep their flight heading in the correct direction. From this, they can realize their wind drift from the reference sight-line. It is certainly possible that a normal means of progression is simply to stop climbing if the wind becomes less favourable. Such a system would also be possible at night using the stars, if the birds are migrating over cloud or the sea. During daylight the sun might be used or, when over the sea, the direction of the waves.

Making a landfall

The problem of where to stop, when the migration flight is ended, is different for each species. A seabird, at sea, may be able to stop and rest almost anywhere. A small warbler, having flown one of its long migration legs, must be able to find a suitable feeding area quickly. Often it will, as a young bird, be arriving in an area where the ecology is completely different from anything it has previously experienced – i.e. different trees and shrubs, a hotter and drier climate and different insects which it must use as food. The problem seems difficult but the solution is rather simple. The birds have evolved to be able to exploit particular sorts of food efficiently. It matters little to them whether the insects are small flies on the lower branches of birch trees in the arctic or small flies, of different species, on the lower branches of thorn scrub in a tropical savannah area. Provided the correct ecological niche is present they are able to exploit it.

A bird flying at 2000 metres will have a very wide field of view and species like waders, ducks and rails will readily be able to identify inland water bodies or coastal lakes and marshes. *In extremis* they will make mistakes and there are many records of unfortunate swans, grebes and divers which have mistaken the shiny wet surface of a road or freeway for a river and landed. Many of these perish in traffic accidents, whilst others, often with severely scratched feet, may be rescued and put on local pools or reservoirs. Newly arrived migrants may have difficulty in exploiting their new food supplies and may often fall victim to new predators – indeed in many parts of the world the tamest birds seen are newly arrived young migrants from the arctic which have no real fear of man.

On their return north in the spring the problem of making a proper landfall is not so acute since they will all have had some experience of the area they are migrating to,

either as young birds or as breeding birds from previous summers. The first migrants to arrive in the spring are, therefore, very likely to be found in the most favoured haunts. For small birds these will usually be sheltered places near water where the early insects congregate and the first flush of spring growth happens.

Birds which are caught over totally inhospitable areas at the end of their flight have several possibilities for survival. They can, if they are over land, simply come to earth and rest hoping that they may find some food. Alternatively they may continue to fly, possibly reducing their speed to one which gives them a maximum duration of flight, hoping to be carried to a better area. If they wish to continue a flight which has been held-up by contrary winds, they may even start to consume part of the muscle tissue as fuel. This sounds appalling but, at the start of their migratory flight they may weigh so much that they need extra muscle to be able to fly efficiently; thus, as their weight decreases during the flight, tissue could be consumed whilst still retaining enough muscle to fly efficiently. Some research workers feel it is likely that many species do this as a matter of routine during the course of their migratory flight whilst others feel it is an emergency move only. Since the reception areas for many migrants with big populations are large, most of them do have a wide target to aim at in the autumn. However, during the spring migration the target area may be equally large but, if they arrive too early, may be cold and inhospitable. The returning spring migrants, therefore, often still have reserves of fat which they are able to use for survival (or even to retrace their steps) if they find there are cold conditions.

Obviously the most important factor to the individual migrant is that it should have enough fuel on board to be able to make the trip. Being able to take advantage of favourable weather and winds to help it on its journey will increase the bird's efficiency but the fuelling is the most important aspect. The full implications of the weight gaining part of the bird's year are dealt with in the chapter entitled Annual Accounts.

Finding the way

To all those who have thought about the problems involved, navigation is the most mysterious aspect of migration. The vast distances involved and the tremendous feats of endurance performed by the birds only serve to underline this most wonderful part of the whole phenomenon – the fact that the birds are able to find their way on long-range flights often to return, year after year, to exactly the same breeding site.

The whole subject is fraught with difficulty. After all you cannot simply ask the birds how they do it but can only painfully piece together information derived from observation or experiment to fill out what is most definitely a very broad canvas. Intensive research over the last decade has shown that birds have a wide variety of 'navigational' options open to them and that, for most species, a single method is seldom, if ever, used to the exclusion of others. An additional complication is that a method of relatively little importance for one species may be that preferred by others.

Before considering the evidence currently available it is as well to consider the problem of 'finding the way' as presented to a migrant bird and the human solutions to similar problems.

The possible solution

If we consider a young bird migrating for the first time which has, until then, only wandered through a relatively small piece of countryside, then finding its wintering area thousands of kilometres away might be accomplished in a number of ways. It might be sensible to migrate with experienced adults who have made the journey before. In this way the young bird will be able to learn where to go so that, in future years, route finding might be possible through remembering where the adult birds had travelled. This kind of strategy is used by many geese and swans and also by some seabirds. In most cases the adult birds travel with their own offspring as a family group, often within a larger group comprising many families from the same part of the breeding area. This would obviously be a very poor strategy for any species which suffers high mortality because the knowledge of the route taken on migration would reside in the memory of the adult birds and, were they to die, the youngsters, upon whom the future of the species would depend, would become lost. There is good evidence within the species adopting this strategy that young birds can make the journey by themselves but that their efficiency is greatly enhanced if they are with their parents.

Another possible course of action would be to fly around at random and hope that a suitable place for wintering was eventually found. However, this would be very inefficient for long-distance flights. A warbler setting off from England or Nova Scotia would have very little chance of making its traditional wintering area using this approach. Interestingly enough random search may have a part to play in finding a particular goal. This is not at all surprising as birds that are in flight have a very wide field of view with the theoretical horizon of more than 150 kilometres, if they are at an altitude of 1000 metres. On a clear moonlit night open water may be visible 50–100 kilometres away to a bird flying at 2000 metres. Many duck watchers and hunters are surprised at how quickly recently flooded areas are visited by ducks. Yet a circling flight up to a good height, taking only a few minutes, will reveal to these birds whether such new habitats are available. However, on the ground, even if you were facing in the correct direction and were only 500 metres away, it is unlikely there would be any visible clues as to the existence of water.

If the young migrant is on its own the simplest means of finding the way is for it to orient its flight in the correct direction and fly. If it carries on for long enough it will get to where it should be and, provided it can recognize that it has arrived, it will have accomplished its first migration. This strategy is probably used by some short-distance migrants but it is certainly not appropriate for long-distance flights on two grounds. The first is that any slight discrepancy between the direction of flight and the correct direction of the wintering area will result in very big errors. For example a five degree mistake in heading would result in a difference of 1000 kilometres at the end of its flight for a Swallow from Europe going to South Africa; this error would be enough to change the destination from the hospitable damp eastern region round Johannesburg to the hot, arid and inhospitable wastes of the Kalahari Desert. Secondly, it is certain that many

species alter their heading one or more times during the course of their migration so as to avoid particularly hazardous areas or, in some cases, to take in favourable stop-over places. If this is the case the young migrant must be able to 'switch' flight direction by some means. This could possibly be accomplished by flying on one heading for a set time and then changing course. It would also be possible for a bird using a set heading (or even a set of different headings) to end up in the correct place by stopping after having flown for a particular length of time.

The final system is the most complicated but also the one which, apart from staying with experienced adults, seems to give the best chance of success. That is to have a goal-oriented system of navigation. With such a system the bird is aware of its position in relation to its destination and should be able to make good any displacements from its 'normal' path: for instance those due to severe weather. Navigation does imply that the bird is capable of doing one of two complex tasks. The first seems very unlikely and is called 'reverse displacement navigation'. This is a system of inertial navigation where the bird plots every move (distance and direction) that it makes in relation to its starting point and 'knows' where its goal is and so modifies its flight to get there. The second task, true navigation, implies that the bird can refer to a grid of co-ordinates provided by two or more celestial or geophysical clues. In special cases, over rather short distances, a single clue could be enough. Following a river to its mouth could easily be all that was necessary for a freshwater breeding duck to reach the sea and its wintering area.

The human equivalents. It is perhaps instructive to look at these possibilities in human terms. The breeding area, where the young bird has been reared, is obviously a close equivalent of the human's home and its surroundings. The migration with the adults may be the same as learning the way to school from one's parents or, possibly, going on holiday with them. Travelling in a particular direction is the same as the street direction: 'Go down the High Street and you can't miss it.' This may be modified by the equally familiar 'Carry on down here and turn left after three miles'. These are all means of dead reckoning. Inertial navigation is at its most sophisticated in nuclear submarines or missiles where a very fast gyroscope provides a means of reference and a computer collects and processes information on all accelerations, decelerations and changes of direction. The final form, with a grid, corresponds to conventional navigation where longitude and latitude are calculated and one's position is plotted on a map. This would normally have involved 'shooting' the sun or stars at an accurately known time to obtain a fix which then, through a set of tables, could be converted into the geographical co-ordinates. Modern methods involving radio beacons (for example, the Decca system) provide very efficient means of obtaining three different grids for extra accuracy.

The human analogy may be taken further. When expressed in human terms it is perfectly obvious that all these methods of finding one's way might be used and that it is very likely that, during the course of any journey, several will be utilized. The captain of a ship will use his sophisticated aids during the course of a voyage when far out at sea; closer to the shore landmarks – headlands, lighthouses, buoys – and a chart will be used whilst when he comes to make fast at the quayside the problem will simply be to manoeuvre the ship rather than discovering where to go. It is easy to see that birds faced with the same sort of navigational problems are likely to use a variety of methods. The other important point is that a resourceful person stranded in a wilderness area without aids will use a wide variety of clues to work out where to go to get to 'safety'. For instance, these might include: using his watch and the sun (a real reason for a conventional dial rather than a digital one) to find south; at night looking for the Pole Star; if the weather is overcast looking at the pattern of vegetation around isolated rocks or bushes (these generally show the effect of shade to the north in the temperate parts of the northern hemisphere); looking for the brightness of dawn in the east – or the dying rays of the sun

in the west; or simply walking downhill (possibly following a stream in the direction of flow in flat country). In much the same way it is most likely that any migrant bird will have a variety of route-finding techniques available should conditions not be appropriate for the most often used method.

The theories and evidence for different navigation methods used by birds

Many years of research have failed to provide the exact answer to the question: 'How do birds navigate?' A great deal is known about some of the means various species use. However, general comparisons from these to other, unrelated species, faced with very different migrations in distant parts of the world, are dangerous. In the following sections various possibilities are examined in several different ways:

1) How can information be used for navigation?
2) How may the bird be able to receive information?
3) What evidence is there that birds are capable of responding?
4) Is there any evidence that alterations to the information produce predictable changes in navigational abilities?
5) Are there any obvious limitations to the method – i.e. can it be used by trans-equatorial migrants, will it provide only 'coarse' information, or may it be seriously affected by local environmental changes?

As the different possibilities are examined some may appear very far-fetched indeed. The possibility of a tiny bird weighing 10 grams – with a brain of only 0.5 grams (the weight of ten standard-sized postage stamps!) – being able to navigate by the stars seems fantastic enough. To suggest that birds can directly sense the magnetic field of the earth might seem even more ridiculous. However, it has not only been proved that they can and do but, in pigeons, a tiny organ has been discovered with crystals of magnetite in it and some 100 million to 1000 million single magnetic dipoles. Even worse, for the doubters, Robin Baker, a British zoologist who has done considerable research in this area, has been able to show that human beings can also perceive the earth's magnetic field and, although they do not realize they are using it, can successfully point to home after a long, twisting and blind-folded journey. With this in mind and remembering the different needs of different species – the sound of waves on a distant shore may be very useful to a seabird but of no practical interest to a migrant which spends its whole life inland in Asia – all the theories should be given serious consideration. It is much easier to propose a theory and to find evidence in its favour than conclusively to disprove one.

Circadian rhythms

The evidence for avian circadian (*circa* = approximate; *dies* = day) rhythms need to be examined, since the first and most obvious sense that birds may use to navigate is sight and, if celestial navigation is being used, many theories depend on the bird having an internal chronometer. A bird in a natural environment will, of course, be influenced by the regular daily rhythms of light and darkness, heat and cold. Such rhythms as daily feeding and roosting have been shown to be directly influenced by local weather: dark overcast conditions cause birds to get up later in the morning and roost earlier in the evening. However, birds kept under artificial conditions can be conditioned into believing that the day is starting earlier (or later) by altering the time when the lights in their aviary come on. If birds which are subjected to a normal 24-hour cycle have their normal cycle changed for instance, dim lighting remaining on all day – then records of their activity show that they maintain a daily rhythm of about a 24-hour 'day' without having any external clues to keep them on time. In fact, the results of hundreds of experiments with birds, mammals, Man and other organisms show that the daily cycle is endogenous (controlled internally) and may free-run, without any external stimuli of, for example, light and dark cycles over months or even years at approximately a 24-hour period.

The exact mechanism used by the bird to run its internal clock has been the subject of intensive research, notably by two German scientists – Aschoff and Gwinner – working at the Max-Plank-Institut in West Germany. It seems likely that several different 'clocks' exist independently and much recent work has been concerned with 'Zeitgebers' – the periodic factors in the environment used by the birds to set their clocks. Simple alteration of light/dark cycles to run with periods of between 22 and 26 hours showed that the bird's circadian rhythm would only respond and entrain with cycles in the range of 23–25 hours. Further experiments also proved that the brain is able to receive information on light/dark cycles directly through the skull and not necessarily through the eyes. Other experiments have proved that exposure to a recorded song of the male in a daily rhythm will entrain the clock of others, for instance, with Serins. Physiological experiments have also indicated that hormonal balance plays an important part and that the pineal gland (sense organ in the brain) is of supreme importance in the circadian rhythm of body temperature and uric acid excretion.

This may seem a very far cry from anything to do with bird navigation but an internal clock of real accuracy would be needed for a bird to undertake full celestial navigation. The Admiralty in Britain were, as recently as 200 years ago, offering substantial prizes to watchmakers for the production of accurate timepieces for navigation. The reason is simple: as the earth rotates once every 24 hours, each hour of the day represents a change of $15°$ longitude. That is a distance of 1800 kilometres at the equator. A difference in time of a single minute represents 30 kilometres. Clearly a very special sort of biological clock would be needed to keep time, day in and day out, to an accuracy of 5 minutes – indicating a navigational capability accurate to about 100 kilometres in the temperate zones. As we shall see time compensation of a sun compass does exist and so the circadian rhythm is used here by migrants as well as for pinpoint celestial navigation.

The result of inaccurate timekeeping is a disaster if the time is being used to set the longitude. Similar mistakes when applied to directional information from a time-com-

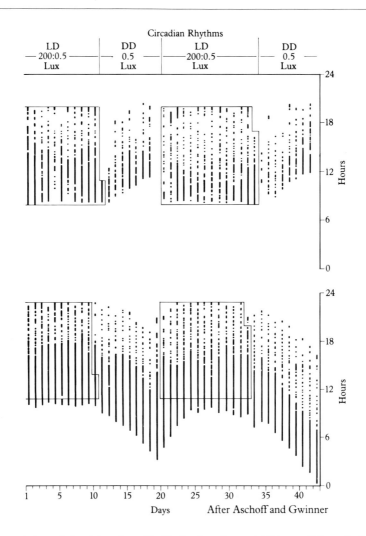

Circadian Rhythms

LD	DD	LD	DD
— 200:0.5 —	0.5	—200:0.5—	0.5
Lux	Lux	Lux	Lux

After Aschoff and Gwinner

These charts record six weeks' activity of two Chaffinches kept in special light conditions. Each line is one day and the boxed areas indicate the brightly lit periods – the left-hand chart is for a bird with an innate rhythm just over 23 hours/day and the other just under 25 hours/day: both return to 24 hours when the lights give them the necessary stimulus. Dark marks indicate times when birds were active.

Mistakes in time keeping

Error in clock	Distance out East to West as longitude error in the latitude of Europe or America	Distance out after 500 kilometre flight due to inaccuracy of direction
1 minute	20 kilometres	2 kilometres
5 minutes	100 ..	11 ..
10 minutes	200 ..	22 ..
30 minutes	600 ..	65 ..

pensated sun compass are not nearly so big:

Visual methods: Landmarks. Anyone who has flown will realize that, from a height of 1000 metres or more, the landscape is set out below like a map. Birds will undoubtedly learn to recognize landmarks if they make such flights but many of the small migrant species seem seldom to fly far from thick cover and thus will not have experienced this 'map-like' view. They may, in some cases, be born and reared close to very obvious features that might be recognizable; for instance, a range of conspicuous mountains, some tens of kilometres away. There is little evidence from experiments that they make much use of such features although some homing pigeon experiments indicated birds making a direct flight to their loft as soon as it was visible. The same experimenters did not discover any evidence that the pigeons were using landmarks on their main flight. Equally it is very obvious that seabirds flying to a small island may be able to see and recognize it at a very great distance. It is also very possible, since many undertake long feeding flights, that they gradually build up memories of a very substantial area and its landmarks.

Robin Baker has attempted to unify all sorts of animal movements under a single

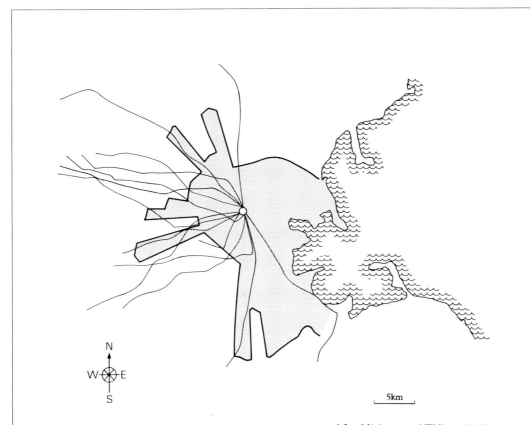

After Michener and Wallcott (1967)

The tracks of homing pigeons returning to the loft (marked by the circle) in Cambridge, Massachusetts. Within the shaded area an observer flying at tree-top height (like the pigeons) would be able to see the tall buildings in the town centre and thus probably recognise where the loft is by these landmarks. Some of the birds were flying fairly directly at the loft anyway but, within the stippled region, a number of the pigeons made a sharp turn towards home – almost certainly because they recognised the tall buildings.

'familiar area' hypothesis. He argues that not only the whole experience of all areas visited and seen by an animal make up its familiar area but that once it has found its way on migration or to a particular feeding, roosting or nesting area, this also becomes a part of its 'familiar area'. The theory has its attractions and one may watch small mammals put into a new situation exploring the area available for them in exactly this manner. Young birds after becoming independent of their parents and before undertaking long-distance migration also disperse and appear to explore areas adjacent to those where they have been reared. The building up of a more or less extensive 'familiar area' with known land-marks and, possibly, some ideas about the suitability of particular parts as potential breeding sites will obviously be a useful facility. It would be even more useful if, as is possible, the bird's long-distance navigational system did not enable it to get back to its preferred breeding area with pin-point accuracy. The theory can be extended to allow for familiarity with areas far from those previously visited if, for example, the bird inherits information about a 'star-map'. Most readers will never have been to Australia but, through maps, it is still a part of their familiar area.

The sun. Countless experiments have shown that birds are able to orient with a fair degree

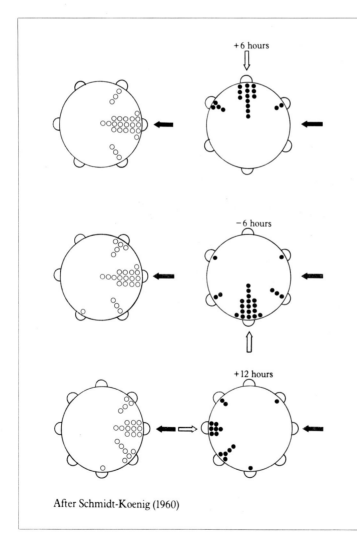

+6 hours

−6 hours

+12 hours

After Schmidt-Koenig (1960)

Experimental results for clock-shifted homing pigeons. Each pair (left and right) represents results obtained at the same time – those on the left from birds experiencing 'normal' time but those on the right had been kept under artificial light conditions so that they believed, (from top to bottom), that the time was six hours earlier, six hours later and twelve hours difference. The results are from trained birds which pecked in a preferred direction and the predictions (marked by white arrows) fully support the time compensated suncompass theory.

of accuracy when they are able to see the sun. This is done by reference to the direction of the sun, not to its altitude and is, of course, time compensated. The classic sort of experiment used to demonstrate this facility involves the bird choosing a switch at the edge of a circular cage for a food reward and then having to return to the centre to collect the reward. The figure shows the results for three pairs of experiments with homing pigeons. The results are exactly as would be predicted. Further experiments have shown that birds from the northern hemisphere that are not trans-equatorial migrants will, when transported south of the equator, orient at 180 degrees to their expected direction. Starlings taken from Germany to north of the Arctic Circle in Sweden were able to maintain their orientation even when the sun was due north of them at local midnight.

The accuracy of the sun-compass cannot really be judged from the scatter of headings shown in the results of the experiments on clock shifting. A neat system of self-training was adopted to see just how accurately a pigeon could read off the sun azimuth (horizontal direction) and altitude. The results were interesting on three counts. Firstly, it showed that both the altitude and azimuth of the sun were measured by the pigeons looking at its shadows, not directly at it. This may be done to protect their eyes from the brightness of the sun but the experiment also showed that it would be possible, theoretically, for the birds to be up to six times more accurate in their measurements by using shadows rather than the sun direct. These findings have not been confirmed in wild birds at large but it would be very easy for flying birds to use shadows on the ground as sightlines. The other two findings were that the accuracy of measurement of azimuth varied between ±3.4 and 5.1 degrees for different birds whilst for altitude it was rather less accurate – between ±8 and 11 degrees. These are not particularly accurate but the azimuth direction, especially if it is constantly updated by the flying bird, would certainly be good enough to enable the pigeons to home accurately if they knew the direction in which their home loft lay.

These results have a very significant effect on the classic theory explaining navigational abilities of birds in daylight, which was pro-

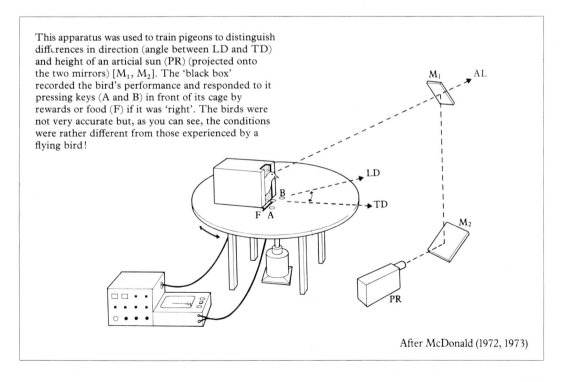

This apparatus was used to train pigeons to distinguish differences in direction (angle between LD and TD) and height of an articial sun (PR) (projected onto the two mirrors) [M_1, M_2]. The 'black box' recorded the bird's performance and responded to it pressing keys (A and B) in front of its cage by rewards or food (F) if it was 'right'. The birds were not very accurate but, as you can see, the conditions were rather different from those experienced by a flying bird!

After McDonald (1972, 1973)

posed by Geoffrey Matthews, author of *Bird Navigation* (1968). This required that migrant birds, in flight, should make precise measurements of the sun's direction, height and movement. The theory was that the angle made by the apparent plane of the sun's movement with the horizontal, which remains constant throughout the year at any latitude, was then used by the bird to provide a north/south reference. Further, the bird would also use the sun's height and movement to calculate a local noon (time of highest altitude of the sun) which, when compared with the bird's internal clock, would enable a comparison east/west to be made with the bird's clock 'longitude'. The performances of birds on migration is so good that such complicated systems of navigation not only seemed reasonable but even necessary to explain it. A further suggestion, again only theoretical since there is, as yet, no experimental evidence to support it, was recently advanced by J. D. Pettigrew working on owls. He suggests that birds may be able to make precise measurements of the sun's position, and even movement, by measuring the shadow cast on the back of the eye by the *pecten oculi* – a heavily pigmented membrane on the optical axis towards the back of the eye.

It is important to remember, when we consider orientation by reference to the sun, that the birds may be able to see the sun in different ways to us. There is no evidence that they are able to see longer wavelength light than humans (that is the infra-red end of the spectrum) but there is some good evidence that they can see shorter wavelength light (the ultra-violet). It has also been shown by experiment that birds can detect the plane of polarization of light – particularly in the ultra-violet part of the spectrum. The plane of polarization of the light in the sky is determined by the direction of the sun and this might be used by birds if the sun were obscured by clouds, had dropped below the horizon or disappeared behind mountains. Direction-finding using the plane of skylight polarization has been demonstrated by Otto von Frisch for bees but its use by birds has yet to be recognized.

One final, obvious factor which could be used for orientation is that the sun rises in the east and sets in the west. Various experiments have shown that a sight of the western sky with the setting sun was important for the orientation of night migrants in experimental conditions. The same may be true for the rising sun for diurnal migrants. Radar observations during migration have shown that many migrants set out shortly after sunset (nocturnal) or shortly after sunrise (diurnal). It is also at such times, when the sun is low in the sky, that the skylight polarization is most obviously oriented.

Stellar clues. The most obvious feature of the night sky is the moon. However, the motion of the moon round the earth is much more complicated than the apparent motion of the sun or stars and it is very unlikely that birds use it for navigation. Similarly, the apparent motion of the planets is complex and unlikely to be helpful. The only remaining visual celestial means of navigation at night is the stars; for several decades it has been known that birds under a starry sky are able to orient in the appropriate direction for their migration. The classic experiments performed by Gustav Kramer, conducted in the 1940s, were the first to use the migratory restlessness (*Zugunruhe*) displayed by caged migrants which should have been migrating at night. He expected that the birds' movements would be oriented in the direction in which they should be flying and was proved correct when he measured this in Blackcaps and Red-backed Shrikes confined in circular cages. These techniques are still used, in many different forms, to this day.

In just the same sort of way that Matthews suggested that birds could perform full navigation using the sun and an accurate internal clock it is also possible to suggest means by which they could use the stars. It would be more complicated since different stars are above the horizon at different times of the year but it would be possible for birds to use the stars – after all human navigators have used them successfully. If this is the case then the birds must be using their internal clock to provide vital information. This would have the inevitable result that time-shifted birds (birds conditioned by an

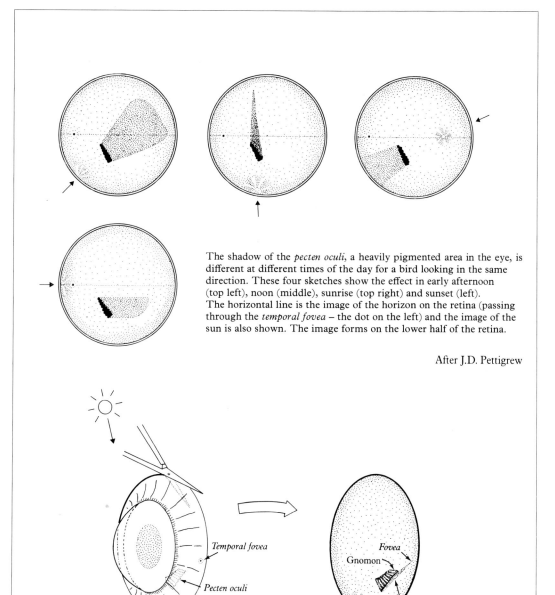

The shadow of the *pecten oculi*, a heavily pigmented area in the eye, is different at different times of the day for a bird looking in the same direction. These four sketches show the effect in early afternoon (top left), noon (middle), sunrise (top right) and sunset (left). The horizontal line is the image of the horizon on the retina (passing through the *temporal fovea* – the dot on the left) and the image of the sun is also shown. The image forms on the lower half of the retina.

After J.D. Pettigrew

Temporal fovea

Pecten oculi

Fovea

Gnomon

Image of Sun

Shadow

If one imagines the sun, at noon, shining on a bird's eye as shown on the left, a special case can be imagined so that, if the front of the eye were removed resultant image of the sun and the *pecten* on the retina could be as shown on the left. Here the image of the *pecten* produces a shadow with a sharp termination actually on the *fovea*. The *fovea* is the area of the retina with the highest sensitivity.

Cages like these were used to establish the effect of magnetic fields on bird navigation.

the stars were unable to orient under normal skys in the planetarium. In their case the star map was certainly not genetically transmitted from parent to chick as had been proposed in some theories. The Indigo Bunting work went further. Birds reared with exposure to a normal night-sky in the planetarium performed normally. However, birds treated in exactly the same way but exposed to the correct planetarium sky **shown without rotation** failed to orientate when later tested under normal conditions. Finally, as the crucial experiment, birds exposed to a falsely rotating sky, around Betelgeuse in Orion, when they were reared behaved as if this were the Pole Star when later tested. This proved that for the Indigo Bunting one way in which the star compass was calibrated was by the birds' experience of the rotation of the stars as it grows up and recognition of the area of least movement. This has a very interesting consequence for it offers a built-in means of compensation by migrant birds for the gradual change in the position of the celestial pole caused by precession. Over a period of

artificial light regime to alter their internal clock) exposed to the normal sky would misdirect themselves by 15 degrees for every hour they were shifted. Also normal birds responding to an artificial sky in a planetarium giving the local sky shifted in time by an hour would show a similar mistake. This is not the case (see diagram opposite). Another experiment which showed the essential difference between time-compensated sun-orientation and time-independent celestial navigation was performed by Matthews investigating the 'nonsense' orientation of the Mallard. This interesting phenomenon leads Mallards released at an unknown place generally to disappear out of sight in the same direction – even when released singly! This and other experiments showed that at least some dissimilar species of birds were certainly using star patterns for orientation, not stellar movement.

In fact the method used by the birds is likely to be very similar to that used by man. The star in the sky currently nearest the celestial pole in the northern hemisphere is Polaris. It can readily be found by referring to the patterns of nearby constellations, even though it is not a particularly bright star. Stephen Emlen showed that his Indigo Buntings were sensitive to alterations made to the planetarium projection within 35 degrees of the Pole Star. What was more, birds reared from the egg without a view of

Testing the navigational abilities of Indigo Buntings in a planetarium in America. The individuals to be tested are in the cages and the star maps projected on the roof may be artificially manipulated to show the sky locally at a different time of the day or year, or to show skies from other parts of the world.

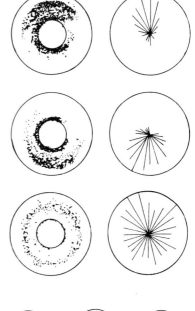

The cutaway above shows an Indigo Bunting in an experimental cone. It is standing on an inked pad within a sloping cone of blotting paper. When it responds to the star pattern by moving in a particular direction it carries ink up that side of the paper producing traces like the three on the right. These are later analysed and interpreted into the familiar diagrams shown on the far right. This simple, self-recording, experimental method is in the very best traditions of experimental science.

Above: The seven pairs of results shown above prove that Indigo Buntings behave the same when put outside and shown the real stars on clear nights (top row) and when shown the local star map in a planetarium (lower row). Each pair of charts are for the same individual bird. This simple checking experiment is crucial to establish the validity of the method.

Right: The six experiments on the right show the directions taken by five lots of clock-shifted Indigo Buntings shown planetarium skies and one (top) not clock-shifted. The bottom plot is of 12-hour shifted birds, the ones above 6-hour shifts and the others 3-hour shifts: in no cases was there a significant deviation from the normal direction (vertical dotted line). Thus, for Indigo Buntings, the star-compass is not time compensated but directly assessed.

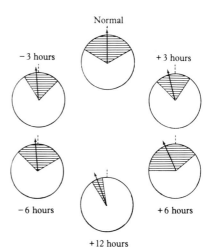

Normal

− 3 hours   + 3 hours

− 6 hours   + 6 hours

+ 12 hours

After S. Emlen (1966, 1967)

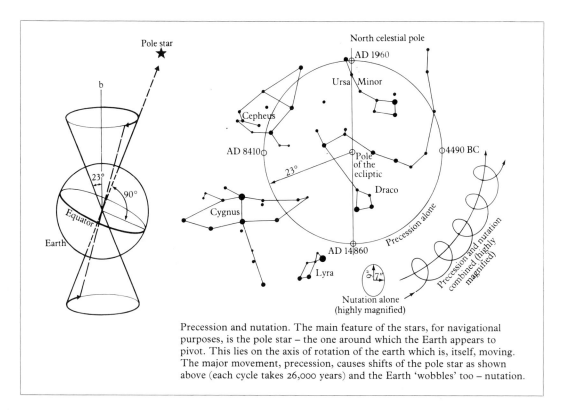

Pole star

b

23°

90°

Equator

Earth

Pole star

North celestial pole

AD 1960

Ursa Minor

Cepheus

AD 8410

23°

Pole of the ecliptic

4490 BC

Draco

Cygnus

Precession alone

Precession and nutation combined (highly magnified)

AD 14 860

Lyra

9″ 7″

Nutation alone
(highly magnified)

Precession and nutation. The main feature of the stars, for navigational
purposes, is the pole star – the one around which the Earth appears to
pivot. This lies on the axis of rotation of the earth which is, itself, moving.
The major movement, precession, causes shifts of the pole star as shown
above (each cycle takes 26,000 years) and the Earth 'wobbles' too – nutation.

26,000 years the celestial pole describes a
circle about 47 degrees in diameter in the sky.

This explanation of how the star compass
works unfortunately relies mostly on detailed
work with the Indigo Bunting which is a
species that does not come near to crossing
the equator in its migrations. Other species
which have provided information confirming
this kind of star compass also keep to the
Northern Hemisphere. Some of the species
tested by E. G. F. Sauer, which he thought
showed evidence for time-compensation,
were trans-equatorial migrants; it could be
that a more complicated form of stellar
navigation remains to be discovered in trans-
equatorial migrants although, as we shall see,
other methods may be used by such birds.

Magnetic clues. The possible use of the
earth's magnetic field by migrating birds has
been discussed for many years. However,
although many experiments were carried
out, occasionally with significant results,
they were never confirmed by other scientists.
Indeed, in one well-known case the same
person, using the same techniques, was

unable to repeat the significant results he had
obtained earlier. This, together with the lack
of physiological evidence of how the bird
might detect magnetic fields, led to the
neglect of magnetism as a source of navigational information for migrating birds by
most ornithologists. Luckily there were some
who were prepared to continue magnetic
orientation experiments and all scientists on
bird navigation owe a great debt to F. W.
Merkel, W. Wiltschko and their team working in Frankfurt.

The evidence that birds can obtain directional information from the earth's magnetism is now overwhelming. It is not, at first
sight, impressive. The crucial results, which
were eventually reproduced by other teams
working on the subject, seem dependent on
the structure of the cages used to test for
migratory restlessness. Cages with radial
perches rather than tangential ones are
needed and the significant results come only
through the pooling of data over several
nights. However, the experiments have shown
that, for the European Robin and some other

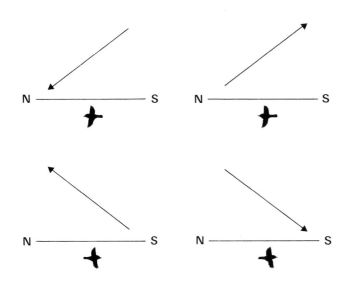

Perception of the earth's magnetic field. Unlike the bar of a magnetic compass, which points to north, birds perceive the earth's magnetic field by the angle it makes with the horizontal – without any reference to the 'sense' (northness or southness) of the field. The birds above are all shown flying towards the acute angle the field makes with the surface of the earth.

After Keeton, W. (1979)

After years of experiments with birds and magnetic fields it was realised that it was important that the birds perched on radial perches as they hopped round the cage and not on ones ranged round the circumference.

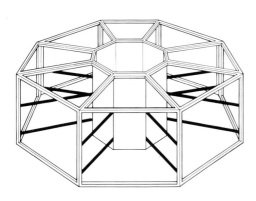

After Merkel and Fromme (1958)

This arrangement of double Helmholtz coils round the experimental cage allows the research worker to adjust the magnetic field, perceived by the experimental bird, to exactly what is needed. Wiltschko's coils were 180 cm in diameter allowing for a good size of recording cage.

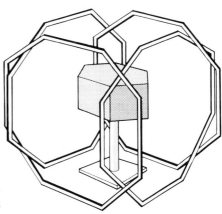

After Wiltschko and Wiltschko (1968)

species, the birds do not orient as if they were following the familiar horizontal compass needle. They are actually using the angle made by the earth's field with the horizontal and aligning themselves as if the acute angle made by the field with the horizontal is north, regardless of the polarity (direction of magnetic field). These cage experiments are by no means the only ones to establish the use of the earth's magnetism by migrant birds and homing pigeons. It gradually became clear that the birds were only able to detect a field if they were moving through it and so, at a stroke, the lack of success which had dogged the many attempts to demonstrate response to magnetic fields in pigeons, strapped into conditioning apparatus, was explained. Furthermore, an astonishing series of experiments by Wiltschko showed that European Robins were only able to respond properly to magnetic fields in the immediate vicinity of the natural terrestrial field of about 0.46 oersteds. Their range was roughly 0.40 to 0.57 oersteds and they were only able to

respond to more or less intense fields after a period of several days.

At the same time as these experiments were being carried on, other research involving pigeons and other birds in flight wearing Helmholtz coils (pairs of identical coils which give a uniform magnetic field) or bar magnets were being performed. Many failed to show the expected effects but eventually it was realized that the sun-compass was a much more powerful immediate stimulus and an effect, as predicted, was seen on the initial orientation capabilities of pigeons released under conditions which simulated a totally overcast day. In a series of experiments near Boston, Massachusetts, C. Walcott released pigeons in an area which had an anomalous field. His birds showed a jumble of tracks until they managed to get out of the anomalous area.

The birds' sensitivity to naturally occurring magnetic storms, often associated with sun-spots, has now been demonstrated. By recording the degree of disturbance at the

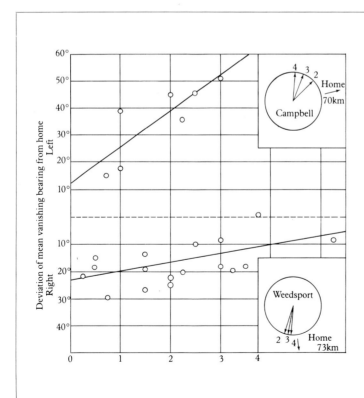

Increasing natural disturbance to the earth's magnetic field (magnetic storms) twist the bird's direction of orientation anti-clockwise. For the upper release point this displaces the birds considerably but, for the lower one, their initial orientation actually becomes more accurate.

After Keeton, W. (1974)

time the pigeons were released, W. T. Keeton was able to show that the birds' initial orientation became twisted counter-clockwise. Indeed, when this change was plotted for one of the release sites for which the birds consistently showed a 'right-of-home' orientation, the magnetic disturbances actually improved the birds' initial direction on release!

The whole story of magnetic clues seems always to lead back to the Frankfurt team. In 1978 they described the 'VW effect': birds which were carried in crates over the engine of their Volkswagen pick-up sometimes showed disturbed orientation; this was ascribed to their proximity, *en route*, to the magnetic field produced by the pick-up generator. This effect was confirmed by others and showed that the pigeons were using the magnetic information available to them as they were being carried to the release site. This might be part of an inertial navigation system which depends on the earth's magnetic field or simply that it takes the birds a long time to 'take a magnetic compass reading'. Anyway, the disturbed clues that they received on the journey masked their real ability.

These results proved that birds do possess a 'sixth sense' which allows them to use the earth's magnetic field for direction finding and many theories and experimental work have been inspired by them. The lack of response to polarity was regarded as puzzling since it would not allow trans-equatorial migrants to distinguish between fields making the same angle with the horizontal (but in different directions) in the Northern and Southern Hemispheres. Indeed this angle can be used to give good information on latitude (see diagram on page 132). However, there is evidence that some species can react to different polarities and the crucial experiments on trans-equatorial migrants have yet to be performed. The magnetic field of the earth has reversed many times during the millions of years that birds have inhabited the planet. It may have been a very reasonable step in natural selection for the species not concerned in trans-equatorial movements to 'drop' measurement of polarity from their

compass system to save confusion at such periods.

Given that birds do have this 'new' sense, two questions immediately arise. How do they detect magnetism? and, as work with many other creatures has shown that they also possess it, does man? A variety of theoretically possible means of detection have been put forward, including one involving the rhodopsin molecules in the eye. All were rendered unlikely by the discovery by E. Gould and C. Walcott of a tiny crystal of a magnetic substance in pigeons, situated in the head between the skull and brain.

Although minute, each bird's crystal probably has millions of magnetic poles. No thoroughly tested physiological explanation of how the detection process works has yet been formulated. The experimental results and a rudimentary knowledge of electromagnetism would point to the direct induction of currents within the bird's nervous system by the movement of its head – and its own magnetic field – through the earth's.

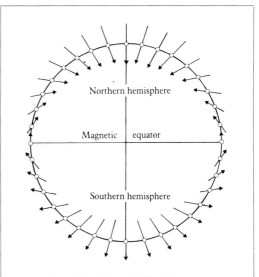

After Wiltschko and Wiltschko (1976)

The strength, angle of dip and direction (polarity) of the earth's magnetic field varies with latitude. The length of the arrow represents the strength – some species are unable to detect the difference in polarity between northern and southern hemisphere.

It is, one would have thought, ridiculous to suppose that man possesses a 'sixth sense'. Robin Baker felt that it should be demonstrated that this was the case and set up a series of tests with his students as the experimental animals. They were taken by minibus to parts of the country unknown to them and then were blind-folded. They were taken from the bus individually and, before their blindfolds were removed, significant numbers were able to point in the direction of 'home'. Later, just like the homing pigeons in America, they were taken on journeys wearing helmets with Helmholtz coils to alter the magnetic field they were experiencing. The results showed conclusively that they were using the earth's magnetic field for their direction finding!

Clues from taste and smell. By far the easiest form of navigation that could be used by birds, or any other organism, would be to have 'home' as a beacon emitting something – light, taste, smell, dye or anything similar – that could be detected by the 'homer', with the simple additional clue that there was more of this signal emanating from that particular direction than elsewhere. The bird could simply follow the signals and would eventually home accurately. Slightly more difficult to envisage is a suite of such clues and the additional ability to detect a certain level of lightness, taste, smell or colour. A simple model shows how this could work (see

diagram below). In birds this is a particularly attractive proposition for the seabirds. Many of them have well-developed olfactory lobes and are probably much better at detecting scents and tastes than most landbirds. The sea varies very considerably in its chemical make-up and most seabirds which come to the surface are able to taste it. They may well be 'tasting their way home'. In addition many of the tube-noses secrete very powerfully smelling oils which they often plaster on their nesting sites. This may be detectable in the air for tens or even hundreds of kilometres. Certainly, if there are individual differences in their odours, it may be possible for the returning member of the pair to find its own burrow at night (many of the 'smelly' species are nocturnal) through this difference.

Even in land-birds with a poorly developed sense of smell the clues provided by odour have been implicated in navigation. F. Papi and his colleagues in Pisa, Italy continued to experiment on this with pigeons, even when it was a much less fashionable subject than the Frankfurt team's attachment to magnetic clues. Their work has shown that, even with the pigeons, olfactory clues have a part to play. Their most successful experiment, where baffles were placed around the loft to displace the winds reaching it, produced unequivocal results (see diagram on page 135). Collaboration with the Cornell

Iso-scent lines – two sources of different scents could provide enough information for proper navigation. The track shows a bird first flying towards the lower source until the strength of the scent is correct for the home site. It then turns along the line giving the same strength for that scent and increasing scent for the upper source until the goal is reached.

I Scent B equals 'home' strength: there is no point in proceeding to any increased concentration.

II Scent A detected for the first time: the bird flies 'up-scent' and along the scent B 'isoline'.

III Home.

Beacon navigation

Source of scent A

Source of scent B

Flightpath of bird flying home

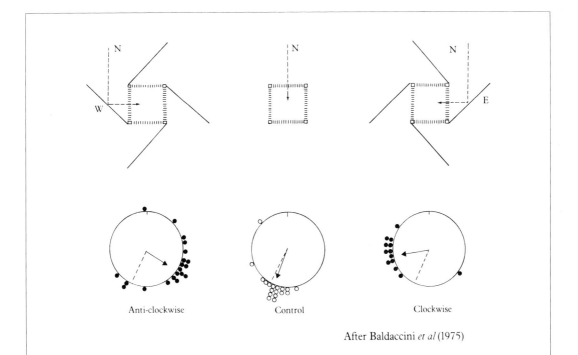

Anti-clockwise   Control   Clockwise

After Baldaccini *et al* (1975)

The famous deflection loft scent experiments. These worked in Italy but were not reproduced in North America!
On the left the deflectors alter the sense of a northern wind to a western one and, on the right a northern
one is turned to an eastern wind. In both cases when released some distance away the pigeons altered their
direction of movement through about 90° in the predicted direction.

team led by Keeton in 1977 led to an agree-
ment to differ as to the significance of odour.
The Cornell pigeons were certainly not so
dependent on it as the Italian ones but the
loft-deflector effect was still apparent.

Aural clues. Many birds call whilst they are
migrating. These calls have usually been
interpreted as contact calls designed to keep
a flock of migrating birds together. However,
playback of calls taped in a balloon has
shown that the echoes from water surfaces
are very clear and it is just possible that
information on wave patterns, and thus wind
direction, may be obtained by such species
when they are flying over water. The possible
use of aural clues does not stop there. Man is
unable to detect infra-sound – that is sound
frequencies below 10 hertz (Hz). Recent conditioning tests on pigeons have shown that
they respond to frequencies as low as 0.06 Hz.
The detailed experiments showed that the
birds were truly 'hearing' these very low

frequency sounds using the inner ear. This
has some very real significance, for the sound
waves produced by the sea breaking on a
rocky shore or the wind striking a mountain
range include infra-sound components which
can be transmitted for hundreds or even
thousands of kilometres with little attenua-
tion. It would, of course, be impossible for the
birds to detect the source of the sound by the
normal method of comparing the signal
received by the two ears – to obtain detectable
differences the ears would have to be hundreds of kilometres apart! On the other hand
a flying bird would possibly be able to
measure the Doppler effect – apparent
shortening of the wavelength as it flew
towards the sound source. This theoretical
use of infrasound has yet to be demonstrated
as being used in a migrant bird.

Although not strictly an aural clue it is
appropriate to mention here another form of
wave which may be very useful to migrating

seabirds. The pattern of waves on the surface of the sea looks complicated and rather random to land-bound observers like man. However, it is anything but unordered and interference patterns, changes in water colour, etc. may offer substantial clues to migrant seabirds. In particular, the influence on wave patterns of isolated oceanic islands can be detected at a range of hundreds of kilometres by the experienced native seafarers of the Polynesian islands. If they can use them there is every chance that the birds are able to benefit as well.

Barometric pressure detection. In areas like the eastern part of North America, where weather systems are rather predictable, it will often help migrant birds to know that they will be assisted by favourable winds if they set out at a particular stage in the weather system. Their response to changes in barometric pressure has therefore been studied with remarkable results. Indeed, not only were they able to detect pressure changes associated with weather systems, but also pressure differences with changes of altitude, and in the experimental pigeons, these could be detected down to 10 metres. This would be most helpful in maintaining height.

Detection of gravitation variations. Recent analyses by T. Larkin and W. T. Keeton showed a significant effect on their homing pigeons, resulting from the state of the lunar month, in the vanishing point of the birds on release. This surprising result has been related to the tidal variations in gravity. If the birds are able to detect these, and there might be other effects varying with the lunar month, then it is possible that birds might be able to detect the tiny variations in gravity with longitude due to the slightly aspherical nature of the earth. Indeed, it is theoretically possible that they might be able to detect the axis of rotation by this means and be able to compare it with magnetic north. The bird's ability to detect magnetic fields seems, to most people, far fetched enough without ascribing to them the ability to detect minute changes in gravity. However, birds have evolved over millions of years and anything which may have conferred the very slightest advantage on an ancestral migrant species is likely to have had a beneficial and selected effect. It is, therefore, possible that detection of gravitational variation has its part to play in the techniques used by birds to find their way.

Coriolis force. The spinning earth exerts an outwards force on objects at its surface – except at the geographic poles – dependent on the distance that part of the surface lies from the axis of rotation. This means that it varies in a different pattern from that of the isolines of the magnetic vertical field – these are centred on the magnetic poles which are over 2000 kilometres from the geographic ones. If the bird was able to detect both then bi-coordinate navigation would be possible over considerable areas. The grid shown on the diagram (see page 137) is not ideal since the area from Chicago through Florida, in North America, has the curves of the two isolines virtually tangential to each other. Unfortunately there is no confirmed experimental evidence to support this hypothesis although some encouraging results were initially reported by H. L. Yeagley. In one ingenious experiment he released trained pigeons from a loft in Pennsylvania near a point in Nebraska where the same two isolines that intersected in Pennsylvania again crossed (a conjugate point). Although he claimed that the birds were oriented towards this conjugate point more recent re-analysis of the original data does not support this.

Putting it all together

There is no doubt whatsoever that much remains to be discovered about the exact means of route-finding practised by migrating birds and by homing pigeons. Magnetism, sun, stars and odour have a part to play with many species and their importance may vary from one species to another. Other means, suspected or not even imagined, may also have a part to play. The relationship between the systems birds are known to use has proved a very fruitful source of research and enables the immediate significance of the different clues available to a bird to be tested.

However, before looking at the evidence of orientation hierarchy, it is salutary to look at the remarkable results, published in 1978, of a single radio-tracking experiment of a homing pigeon. The bird was fitted with frosted contact lenses which effectively fogged everything in view further than 5 metres away so that the bird was unable to recognize objects at all. This particular bird was released about 30 kilometres away from its home loft. Not only did the bird set off in the correct direction – as many birds so treated have done for various experimenters – but it obviously realized when it had reached the vicinity of the loft and actually circled the locality – within a few thousand metres of home – several times. This bird could not see the loft and yet possessed some means of knowing it was in the correct area with a precision that may only readily be explained if the bird possessed a fine bi-coordinate navigational system. Even olfactory clues seem unlikely in this case for the pattern of search would be very different if smell were the major stimulus. The necessary precision in 'reading' the magnetic clues, let alone any gravitational ones, would seem to be beyond the bird's capabilities and yet the results show the bird was able to perform this amazing feat.

The Major Compasses. Three compasses have so far been discovered – sun, stellar and magnetic – and they may all impinge on each other. There is growing evidence that birds find it very difficult to take an instant reading from the magnetic compass. An experiment with pigeons tested after being reared from the egg with their 'photo-period' shifted by six hours (backwards) formed a very interesting test. These pigeons were exercised in the common daylight – real afternoon, their morning – and then tested as to their homing ability. It was just as good as normal. However, when the experimental birds were allowed to regain the normal dark/light cycle they performed like birds clock-shifted by six hours forwards. In later releases they had been able to readjust their sun-compass and performed accurately again. This provides interesting evidence that the magnetic compass is used to calibrate the sun-compass over a period of days. Thus the birds rely on the magnetic compass but the sun compass is being used for the instantaneous assessment of direction.

Wiltschko has also produced some convincing evidence that birds may use their magnetic compass to calibrate the star compass. This included getting birds to orientate when exposed to an artificial star map first shown with a normal magnetic field. The series of experiments on Old World warblers, including trans-equatorial migrants, showed

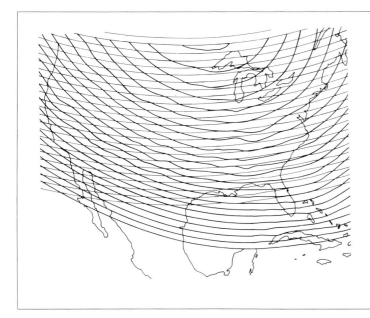

A bi-coordinate grid formed by isolines of the magnetic vertical and Coriolis forces. The flatter, regular lines on the upper part of the map are the Coriolis force isolines – this is the effect of the rotation of the earth which varies according to how far from the axis of rotation a body is situated. This theoretical navigational system has not been supported by any rigorous experimental evidence.

After Yeagley (1947)

A frosted contact lens being fitted to a trained pigeon. This clouds their distant view of the world but does not impair their ability to home to their own loft. The lens is made of gelatin and dissolves away after a few hours if the bird is not recaught.

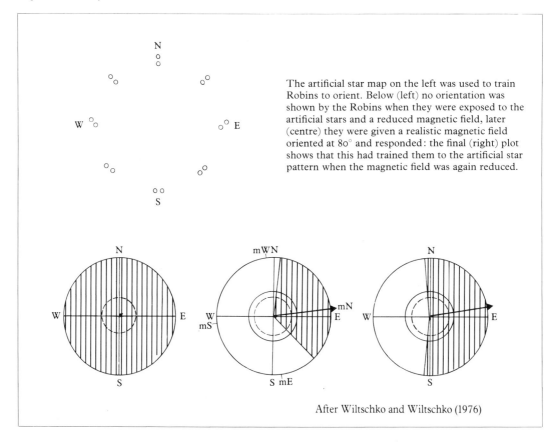

The artificial star map on the left was used to train Robins to orient. Below (left) no orientation was shown by the Robins when they were exposed to the artificial stars and a reduced magnetic field, later (centre) they were given a realistic magnetic field oriented at 80° and responded: the final (right) plot shows that this had trained them to the artificial star pattern when the magnetic field was again reduced.

After Wiltschko and Wiltschko (1976)

that they may have been re-calibrating their star compass rather frequently. This would be very useful for birds crossing the equator where the earth's magnetic field is horizontal and thus not available if they are only able to detect inclination. Therefore all three major compasses are likely to be used together by birds seeking directional information. This is a very sensible arrangement since taking information from as wide a range of sources and integrating it is likely to produce the best possible estimate. In any case, the particular conditions best suiting any one of the direction finding methods may not be available when needed.

A particularly clear illustration of the possibilities provided to birds for compass orientation when one compass is unusable was provided by the 'no-sun-pigeon' experiment. These were birds reared indoors and only exercised on overcast days. Taken for their first homing flight, also under totally overcast conditions, they oriented well. Such conditions cause normal first-flight birds to leave randomly. It is perhaps even more interesting that this should be so since

normal first-flight birds, wearing magnet bars to disrupt their magnetic compass, are also unable to orientate, even in sunny conditions, when birds, wearing brass bars rather than magnets, perform well. Thus, in the case of normal first-flight birds, both sun and magnetic compasses are being used together in a way quite unlike that used by experienced birds – they are able to perform well provided that one or other of the clues is available.

Local effects. The drastic effects of magnetic anomalies have already been mentioned, as has the apparent lack of interest in local landmarks for route finding. It is, however, very well established that local physical features of the environment have their effect on migrating birds under certain circumstances. A particularly clear example is shown for a group of pigeons released by G. Wagner in Switzerland. They did not pick the direction of 'home' and then fly in exactly that direction without any regard for the topography of the area. Their initial 10 minutes looked fairly disorganized but the birds were in fact getting across some very formidable obstacles. Once over them, their track remained very close to

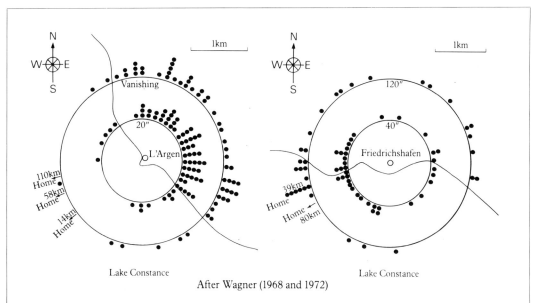

After Wagner (1968 and 1972)

When released beside Lake Constance Homing Pigeons preferred not to fly across the water. On the left the home directions (arrows) were directly out over the lake and most birds after 20 seconds and on vanishing (inner and outer circles) were heading inland. On the right more crossed the water (circles at 40 and 120 seconds).

the home direction. Wagner was also able to show that his pigeons were fairly reluctant to cross large bodies of water immediately after release although, once they were established on the home direction, they were not so deterred. It seems that their confidence in the direction taken is reinforced during the course of the flight. This supports the importance of the magnetic compass which seems to require some time to take a reading from and also to require movement on the part of the bird.

Local effects do have a very great importance since they also establish the 'home' to which the bird will eventually try to return. With pigeons this is the loft but in the case of wild migrants there may be a series of 'homes' which need to be established. This starts with the place of rearing. This, the 'breeding-site' home, certainly seems to be established over a short period, either whilst the young are still in the nest or shortly after fledging. Displacement of hand-reared Pied Flycatchers during this period has led to the establishment of new breeding populations. Stephen Emlen, working with Indigo Buntings, established the existence of such a period in the setting up of the bird's star compass. It is also known to be the period when homing pigeons become attached to their home lofts.

The establishment of the winter quarters is obviously of great importance to migrant birds as this will enable them to have a goal to aim for. Indeed, it means that the wintering area is part of the 'familiar area' for each adult bird. In a classic series of experiments, A. C. Perdeck displaced migrating Starlings from Holland south to Switzerland and analysed the subsequent ringing returns reported. The adult birds invariably started to make their way north-west to the area where they had presumably wintered in earlier years. The young birds carried on to the south-west and ended up in a wintering area which their part of the European breeding population would seldom, if ever, reach. These results also offer an interesting insight into the means by which the young birds came to finish their migration. It was certainly not as a result of reaching an aimed for goal using bi-coordinate navigation. However, tempting though it is

to explain it as simply 'flying for long enough to get there', it is certainly possible that the final wintering area shared another characteristic with the expected location. For instance, it could be that it was the same distance west, as measured by internal clock and local maximum sun azimuth.

One exciting experiment which throws light on the attainment of the winter quarters in birds which are known to have changed direction during their migration was provided by E. Gwinner and W. Wiltschko whilst working with Garden Warblers under artificial conditions. Their results (see diagram on page 142), indicating a change in direction of migration, coincided with the time of year

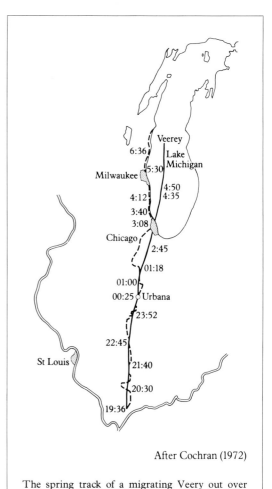

After Cochran (1972)

The spring track of a migrating Veery out over Lake Michigan (solid line) and that taken by the car monitoring its radio transmitter (dashed line).

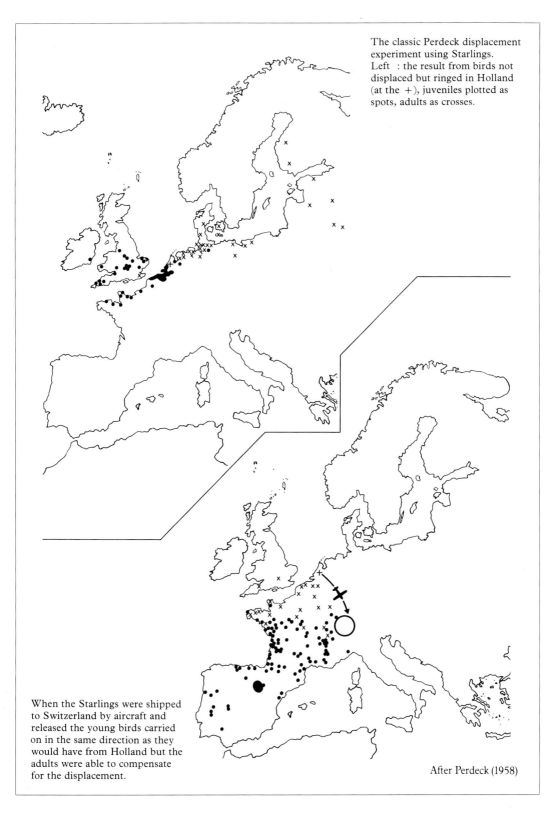

The classic Perdeck displacement experiment using Starlings. Left : the result from birds not displaced but ringed in Holland (at the +), juveniles plotted as spots, adults as crosses.

When the Starlings were shipped to Switzerland by aircraft and released the young birds carried on in the same direction as they would have from Holland but the adults were able to compensate for the displacement.

After Perdeck (1958)

141

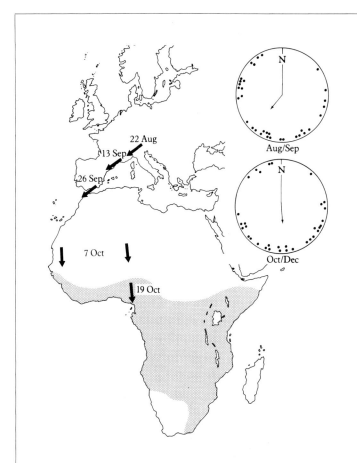

The map shows how the route taken by Garden Warblers on their autumn migration bends round from south-west in Europe to due south in Africa. On the right are the results for birds tested in August and September (top) and October and November (below). Although scattered there is clear orientation to the south-west in the top results and due south below.

22 Aug

13 Sep

26 Sep

7 Oct

19 Oct

Aug/Sep

Oct/Dec

After Gwinner and Wiltschko (1978)

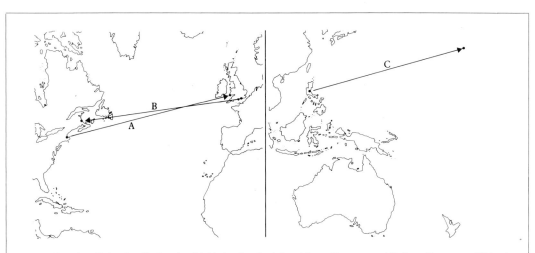

Record homing flights by displaced wild birds. On the left a Manx Shearwater (A) from Boston to Wales in 12½ days and a Leach's Petrel (B) from England to Canada in less than 14 days (distances 4500 and 4800 kms). On the right a Laysan Albatross (C) from the Philippines to Midway Island (6600 kms) in 32 days.

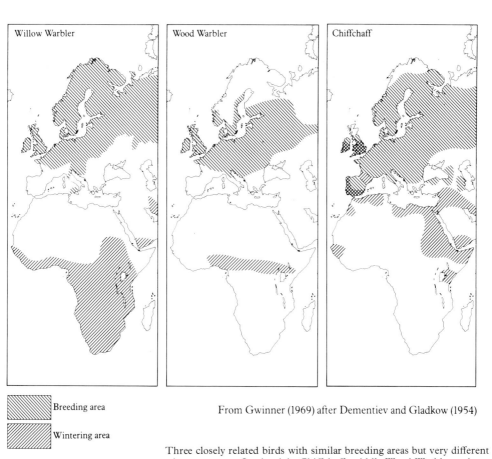

Breeding area	
Wintering area	

From Gwinner (1969) after Dementiev and Gladkow (1954)

Three closely related birds with similar breeding areas but very different winter quarters. On the right Chiffchaff, middle Wood Warbler and to the left Willow Warbler. In experiments the amount of migratory restlessness shown corresponds well with the different distances the three species have to cover.

when the natural part of the Garden Warbler population should have been moving in different directions. The remarkable point is that the experimental birds were being kept in 12 hour dark/light cycles, at constant temperature and without sight of the sky. The navigational clue that they were assumed to be using was the natural magnetic field at their aviaries.

Gwinner's work with the three most common western European *Phylloscopus* warblers – Willow Warbler, Wood Warbler and Chiffchaff – has also shown that the amount of migratory restlessness of the three species corresponds very encouragingly with the different distances they cover on migration. As such this provides more evidence of the vector navigation shown by Perdeck's young Starlings. With the *Phylloscopus* warblers such a strategy may be very important since so few of the birds survive to make the journey twice.

Homing. Much of the experimental work on navigation has been carried out on homing pigeons. This is a matter of convenience only since the ancestor stock of Rock Doves, from which they are derived, are not a naturally migratory group of birds. This does not invalidate the work done but it does mean that there may be other means of navigation much better developed in more natural migrants. After all, the trained homing pigeon is being called upon to perform a task directed towards the same goal every time it is released.

Homing is analogous to the migrant bird setting out for the known breeding site or, in the other direction, for a known wintering area. Artificial displacement of wild migrant birds has also been practised many times in the past. In many cases few of the birds have made the journey and those that have done so have completed it rather slowly. In most cases the birds have been removed during the breeding season when they are not physiologically prepared for migration so poor results might be expected. The successes have often been spectacular, particularly with seabirds which are more likely to forage over great distances, even during the breeding season. (For some well-known examples see the map on page 142.)

Displacement experiments in the winter

After Mewaldt (1963, 1964)

Migrant sparrows from Alaska caught in the winter at San José and released at Baton Rouge returned to San José the next winter. When these birds were taken to Laurel the next winter six of the 22 got back to San José subsequently! In all cases the homing birds will have made the journey back to the breeding grounds in Alaska before returning to the winter area at San José.

are less numerous. Some work with *Zonotrichia* sparrows wintering in California has shown interesting results. Over 10 per cent of 1000 birds displaced by distances of between 5 and 260 kilometres homed successfully but even more remained within the vicinity of the release site. Most of the returning birds were experienced adults whilst those that stayed at the release site were mostly young birds. Further removals over much longer distances also provided successful returns but removal to Asia – over 9000 kilometres away by the Great Circle route – proved too much and none returned.

Back to the individual bird. At the moment we have a fair idea of some of the clues that an individual bird on migration is using to make its journey. However, there are still many questions unanswered which need to be asked of real migrant birds and preferably in a natural situation rather than in the laboratory. Bird ringing cannot provide the detailed evidence needed and so various radio- and radar-tracking studies have been undertaken. Three of these deserve a mention. In a few years they are likely to be looked upon as the crude pioneering studies in a whole new area of research, especially as radio-tracking transmitters get smaller and tracking via satellites becomes a practical proposition.

The first study is of a radio-transmitting Veery (North American thrush) tracked for ten hours one night. This bird was obviously much better able to make the trip than the tracker's car – the latter was forced to use roads and skirt the water of Lake Michigan. If the continuing evidence provided by millions of birds that they can and do migrate successfully, then this bird, actively sending out a signal from the tiny transmitter on its back, should remind us how efficient they are. Many more birds have been tracked individually using highly sophisticated equipment.

The tremendous work done off the eastern seaboard of North America by T. C. and J. M. Williams and their team serves as an excellent example. By using their radar to track individual birds they were able to plot the tracks of birds over Bermuda. These showed a steady orientation in the preferred

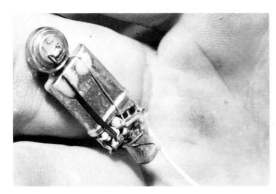

A tiny radio transmitter (*above*) can be attached to a migrating bird enabling the researcher to track its flight path accurately. Photographs below and right show the transmitter being fitted to a thrush and in position on a bunting.

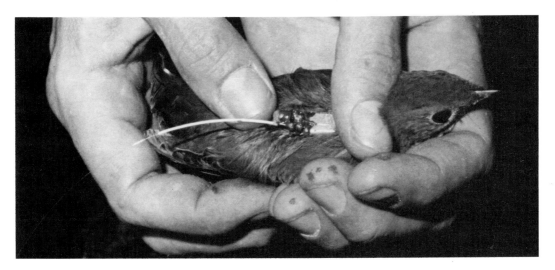

direction with only one bird showing any significant change in direction over tracked flights of from 15 to 60 kilometres. These birds, over 1000 kilometres out in the Atlantic, clearly knew what they were doing. In an extension of their study they used the same technique at seven other places from Nova Scotia in the north to Tobago in the south and produced compass rose plots of the directions taken by the small migrants they were able to track. The south-easterly direction (see map on page xxx) preferred by many birds at the four northern sites would actually take the migrants down the middle of the

South Atlantic if they retained the heading. In fact, the turn in direction to the west may be deliberate or may come about through the birds reaching the area of the northeast trades at about 25°N. Whichever it is they are able to make a landfall in the eastern West Indies or South America some 80 hours or more after setting off from Nova Scotia or Cape Cod.

The final radar tracking experiment concerns the use of freshly caught, and therefore properly identified, birds taken aloft in a cardboard box suspended under a balloon. By remote control the box is opened and

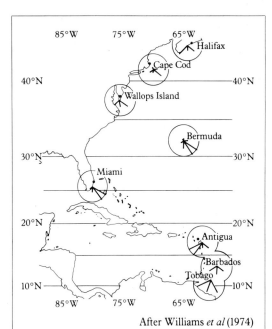

After Williams *et al* (1974)

The trans-Atlantic radar migration watch in fall 1973 clearly shows birds moving south-east from the North American coast and over Bermuda but turning to be heading south-west when they make their southern landfall.

Stephen Emlen and his team were able to track known individual wild birds released at night to discover their reaction.

The State of our Knowledge. Each migrant species may use all or any of three known compasses – magnetic, sun and stellar – to determine direction. Other environmental features which may be used in route-finding include familiar landmarks, odour, taste, infra-sound and even changes in gravitation or Coriolis force. There seem to be many instances where birds perform better than should be expected from our present state of knowledge. All current theories which seek to explain this through a system of bi- or poly-coordinate mapping remain only theories. The only full experimentally verified explanation of attainment of a wintering area remains the vector navigation – flying in a particular direction for a set distance – of young Starlings. Such a method could explain many 'first-migration' situations but very precise skills in navigation, as yet unexplained, are used by homing birds and second-time and subsequent journey migrants.

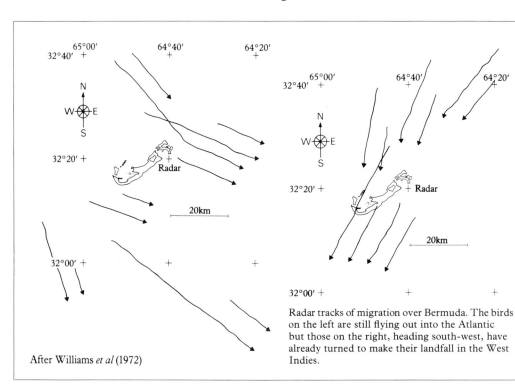

After Williams *et al* (1972)

Radar tracks of migration over Bermuda. The birds on the left are still flying out into the Atlantic but those on the right, heading south-west, have already turned to make their landfall in the West Indies.

The annual
accounts

An important part of the lives of migrant birds is formed by their migrations but, for the species to survive, all the other aspects of the annual cycle must be integrated within its year. For the species to survive, breeding – the provision of future generations – is obviously of the greatest importance but, for the survival of the individual bird, the moult – renewal of feathers – is probably of the highest priority. Both these activities involve alterations in the individual birds' physiology as does migration, particularly the build-up of fat deposits as fuel for the journey. The integration of these activities into the birds' year is the subject of this chapter.

In general terms the annual cycle settles down to a steady rhythm for a small bird after it reaches the age of about 18 months, provided that it manages to survive. Most small birds have a very poor survival rate which is balanced by the relatively large broods of youngsters that the breeding birds are able to rear. The diagram shows the main parts of this 18-month period for a single-brooded, long-distance migrant from northern latitudes to the tropics and breeding in its first summer. At first sight this seems to be a neat arrangement into which the different species can readily slot themselves. However, when

we come to look at particular species and consider the amount of time needed for each activity it is easy to see that many of them are forced to keep to a very strict timetable.

The situation is rather different for a large seabird which may not breed until it is several years old. With such birds survival rate is high and the breeding rate low. Pre-breeding birds, that is those not old enough to breed, often migrate further than the breeding birds and may remain in the winter quarters during their first summer or two. Later they use the summers to become used to the areas in which they will eventually breed to establish breeding territories – even occupying burrows and building nests – and to find a mate. Many species use highly skilled fishing techniques which they perfect over their first year of life, often being taught by their parents for weeks or even months after they have fledged, but it seems possible that learning where to fish during the breeding season may also be very important. Feeding studies of breeding birds of some species have shown that even experienced breeding adults may have difficulty in finding food for their chick. In many areas where human fishing activities have altered over the last few decades fish stocks have changed out of all

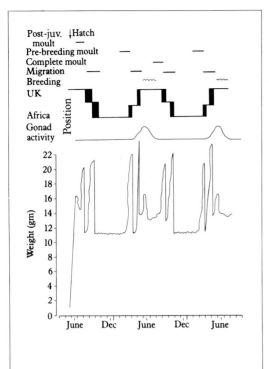

Annual activity pattern of an Old World warbler or chat breeding in Europe and wintering in Africa. The major weight increases correspond with the major migration journeys – two at each season.

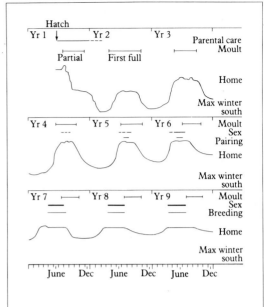

The first nine years of activity for a large seabird. Weight fluctuations (not shown) much less than for a small migrant: maxima on fledging and during early part of each full breeding attempt.

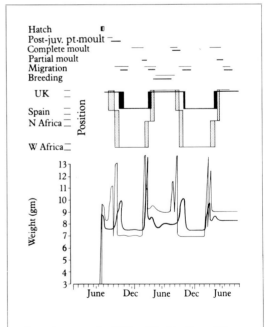

Annual patterns for Willow Warbler (long distance migrant, thin lines and stipple) and Chiffchaff (short distance, thick line and solid) compared over first two years of life.

recognition. Sometimes this imposes extra pressures on the birds but sometimes it helps them; for example, Gannets and Fulmars have expanded their populations greatly over the last 50 years in the eastern Atlantic.

Circannual cycles

Just as birds possess an internal clock set-up to produce circadian rhythms, as described in the previous chapter, they also have a longer term clock which is at least partially in control of their annual cycle. In fact recent research has shown that there may be several different annual clocks which affect different aspects of the annual cycle. There are some activities which are much less influenced by the birds' own clock and much more susceptible to external stimuli. The laboratory investigations of these cycles have generally chosen easily measured aspects of the birds' biology to investigate. These include the birds' weight, its state of moult and its bouts of migratory restlessness or *Zugunruhe* (the nocturnal activity indicating that it was in a physiological state to migrate). There are other major aspects of the annual cycle which are not so easily measured without harming the bird: for instance the development and regression of the males' testes and females' ovaries at the start and end of the breeding season. The experimental demonstration that these cycles are controlled within the bird and are not simply brought on by the annual changes to the environment has now been accomplished by many groups. German physiologists, particularly Eberhard Gwinner working at the Max-Planck-Institut, have been very successful at running experiments lasting two or more years to follow through the cycles of birds in aviaries. The sort of results obtained are illustrated by two *Phylloscopus* warblers, the Willow Warbler and the Chiffchaff.

This species pair is of particular interest. Both breed in the same latitudes but the Willow Warbler is a long-distance migrant, reaching as far south as 30°S, whilst the Chiffchaff travels much less far with many, possibly the majority, wintering north of the Sahara. Those that do cross the Sahara never regularly cross the equator. Since they have to undertake such a long migration the Willow Warblers begin to move much earlier in the autumn than the Chiffchaffs and they also arrive back in the breeding quarters later in the spring. The Willow Warbler is also unique, as far as is known, in that both the post-nuptial moult in the breeding area and the pre-nuptial moult in Africa are complete and thus involve not only the body feathers but also the wings and tail. Gwinner's work with captive birds involved keeping control individuals in cages but exposed to the changing light-dark cycles of Germany. The real experimental birds were transferred to a static photo-period (most often 12 hours light and 12 hours dark) and kept at a constant temperature. The different timing of the start and duration of the measurable aspects of the annual cycle can then be compared. Some of the results are summarized in the diagram (see page 150); in general the Willow

Warblers keep more accurately to the real annual cycle under the constant conditions than the Chiffchaffs. Both species show differences from the cycles that would be expected in the wild for, even for the birds kept under natural light regimes, life in the aviary would be very artificial with no chance of actual migration or proper territorial or breeding behaviour. Nonetheless the similarities between the experimental and control birds are striking.

The control birds also produced results which clearly demonstrated that the first autumn migration started earlier for Willow Warblers than for Chiffchaffs, even when they were transferred to constant light-dark conditions. In an interesting additional experiment Gwinner obtained some chicks of the northern race of Willow Warblers, *acredula* from Sweden, to compare with the local German *trochilus*. The northern birds, bred above the Arctic Circle, had hatched later than the German ones for there the breeding season starts later. However, because the summer is shorter they are also forced to migrate earlier and so, if these timings were endogenous, the northern birds kept in the German aviary should get through their post-juvenile moult faster and start to show *Zugenruhe* earlier too. The results were conclusive and showed not only that the northern birds got through their post-juvenile moult faster than the southern ones but that they started to show strong *Zugenruhe* even before it was finished. This last feature was particularly apparent in the experimental birds given a long day (18 hours) and short night – the sort of conditions appertaining to their breeding grounds in late summer.

This fascinating series of experiments also yielded very interesting data on sibling birds (birds taken from the same nest). Not only were their parents the same but their treatments were also identical. Gwinner showed that there was quite a high degree of variability between the sibling birds but that this was much lower (about half as much) in the Willow Warblers as in the Chiffchaffs. Some variation is to be expected as, though their genes are shared with their parents, the

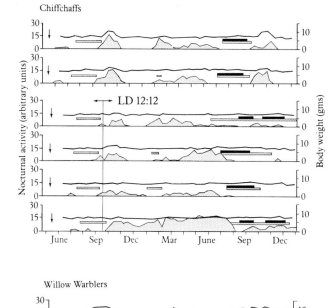

Chiffchaffs

Nocturnal activity (arbitrary units)

Body weight (gms)

LD 12:12

June　Sep　Dec　Mar　June　Sep　Dec

Activity patterns of six Chiffchaffs reared in captivity. The top two birds were exposed to natural fluctuations in day-length but the lower four birds experienced a constant 12 hour light-dark regime and constant temperature from mid-September. The body weight (graph), moult (bars – solid of main flight feathers, stipple of body) and migratory activity (area of stipple under graph) are plotted and show similar patterns but some 'drift'.

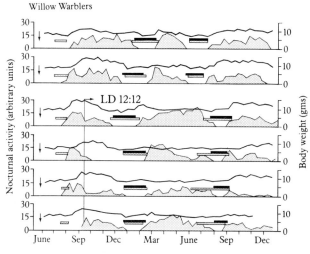

Willow Warblers

Nocturnal activity (arbitrary units)

Body weight (gms)

LD 12:12

June　Sep　Dec　Mar　June　Sep　Dec

These activity patterns for Willow Warblers reared in captivity are set out as those for Chiffchaffs above. The Willow Warbler patterns continue in the constant light-dark regime with a much stronger 'natural' pattern than shown by the less migratory Chiffchaff. Tremendous avicultural skill is needed to rear these birds and keep them healthy for such long periods under artificial conditions.

After Gwinner, E. (1972)

individual birds will be unique. For both the species a spreading of birds from the same nest to slightly different wintering areas would also have its advantages. However, because of their different migration strategies, it is also possible to explain the difference in intrafamily variability. Chiffchaffs move less far than Willow Warblers and respond much more to differences in the timing of spring weather.

All over their range the migration times in the spring may be different by ten days or a fortnight between warm and cold seasons. The Willow Warbler, on the other hand, goes much further and reaches tropical regions where there may be very little seasonal variation in weather or day-length to provide extra clues for the timing of the return migration. They will, therefore, be much more reliant on the circannual clock. This fact is supported by the very regular timing of the spring return to the breeding quarters which rarely varies by more than a day or two, regardless of local weather conditions.

The annual cycle in natural populations

Experimental investigations such as Gwinner's have provided very important and detailed data on individual birds. Other surveys are able to provide extensive information about whole populations of birds. These are usually the results of large numbers of birdwatchers making specialist observations and then pooling their results. For example, the British Trust for Ornithology has gathered, over three decades, all sorts of information about British Lesser Whitethroats which can be put together to give a very clear picture of how they behave in Britain. Arrival and departure dates can be obtained from the records of the south coast bird observatories. The nest record cards give details of the breeding cycle, including number of breeding attempts and the time spent nest-building. Bird ringers who fill in moult records are providing data for the estimation of the timing and duration of moult. The ringers are also able to provide weight information throughout the bird's time in Britain. In the case of the Lesser Whitethroat the ringing recoveries also provide information about the timing of the bird's passage through its intermediate staging posts in both autumn and spring. Local ringers were able to provide some weight data from these sites and a British ringer working in the winter quarters (Ethiopia) was able to complete the picture with winter weights. All this data is displayed in the diagram (see page 152).

The most striking point is how very crowded the birds' time in Britain is, especially when compared with the months in Ethiopia before weight is gained for the spring passage. A human work study consultant would surely recommend that the complete autumn moult should be shifted from the autumn in Britain to the winter in Ethiopia to enable the crowded summer schedule to be spread out. It seems likely that the schedule has evolved like this because the summer breeding activities are most likely to wear the adults' feathers and because the late summer, after the young have fledged, is a time when there is a great deal of food about. The adults can take advantage of this, completing their moult at the same time as putting on their fat for the first leg of migration and migrate with new feathers in good condition.

The same sort of pattern can be seen in many small land-birds which migrate from temperate northern regions to winter in tropical or sub-tropical areas. The detailed timing of the birds' cycle will depend on the pressures imposed on it by the type of environment it lives in and the weather conditions which they experience. For example, the Siberian Lesser Whitethroat populations of the race *blythi*, which breed far from the warming influence of the Atlantic, may not return until late May or even June. By late August these populations are already moving south and all have left by early September. It seems that they still undertake their complete moult on the breeding grounds. They are certainly incapable of raising two broods unless the nests of the second clutches are being attended by adults which are both undergoing their complete moult and starting to put on weight for migration. This might seem an impossible task but, in such climatic situations, food often becomes very abundant for a short period in the late summer and it is just possible that this is what does happen. Certainly the Lesser Whitethroat population of the area are able to persist even though their summer season is so short.

The passerine with the most restricted summer is the Snow Bunting which breeds in the high arctic. For them the choice of territory is of paramount importance; studies have shown that the males arrive near the breeding grounds in eastern Greenland in March when they are still in the grip of winter. The males remain gregarious until the first signs of the spring thaw and then disperse to the breeding areas, starting to defend aggressively territories which often, when they first occupy them, are completely frozen. The females arrive in April but the breeding attempt does not start until the thaw is well under way; the hatching of the nestlings must occur when the insects become abundant in the summer. When this does happen in July and August, the food is easily obtained and the chicks are quickly reared.

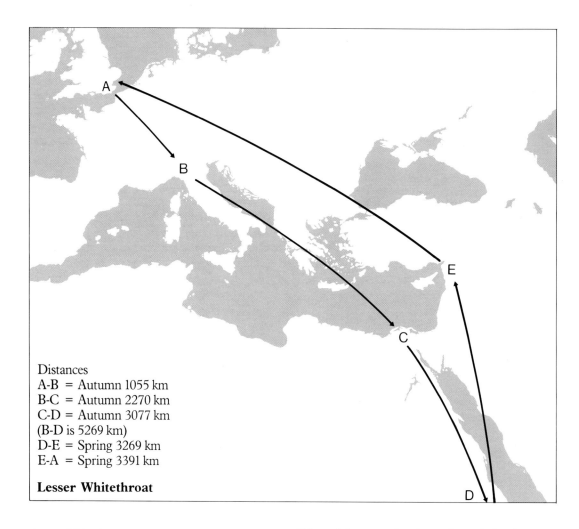

Distances
A-B = Autumn 1055 km
B-C = Autumn 2270 km
C-D = Autumn 3077 km
(B-D is 5269 km)
D-E = Spring 3269 km
E-A = Spring 3391 km

Lesser Whitethroat

Lesser Whitethroat

This small *Sylvia* warbler breeds in scrubby areas from Britain in the west to halfway across Asia. The British population (A) migrates southeast and rounds the eastern end of the Mediterranean. There are ringing recoveries in the autumn from two areas – northern Italy (marked B on the map) and around the mouth of the Nile in Egypt (C). None have been found in the wintering area (D). In the spring the few recoveries come from Asia Minor and Cyprus (marked E on the map). These probably represent the different stages on migration of the British population of this small bird. The distances involved, shown on the map, are certainly well within the theoretical capability of a small bird using stored fat for fuel. A trip of 1000 kilometres would require about 20% of the fat free weight to be taken on as fuel, 2270 kilometres just under 50% and 3400 kilometres just over 80%.

The adults completely replace their plumage immediately after the young fledge and moult so quickly that they generally become flight-less for a few days. At the same time they are putting on weight for migration and will have moved southwards from the bleakest breeding areas before the end of August.

With most of the waders the annual cycle of the northern breeding populations is subject to the same sort of climatic constraints as the Snow Buntings. But, since their chicks are often not able to develop with the speed of the passerines, the cold of the autumn arrives before the adults are able to complete their moult. This has meant that many of them migrate southwards before moulting and several species have special stop-over points where moulting takes place over a few weeks in August or September. Some species split: the young birds migrate at different times from the adults, since they, the youngsters, do not have to replace all their feathers. Detailed ringing studies in western Europe have shown that many waders are mobile even during the winter and the centre of gravity of whole populations shifts month by month.

The diagram (page 154) shows the complex series of movements and massive changes in weight undergone by the high arctic breeding populations of Knot from northwestern Greenland and northern Canada. These complicated autumn and winter migration patterns have probably evolved in response to the changing availability of food over the western European estuaries through the winter. The Waddenzee area and the Dutch estuaries experience much colder winter weather than the British sites. It, therefore, probably pays to exploit them early in the season leaving the milder areas in Britain for mid-winter and, in the mildest western parts, for weight gain before the spring migratory flight.

Special moult migrations are also a feature of the annual cycle of many of the ducks and geese. However, the geese and swans have a special attribute shared with few other groups. They tend to stay together as family parties throughout the migration, during autumn, winter and the spring return. This can obviously be of great benefit to the young birds and, by increasing their chances of survival, also ensures that their parent's genes are more likely to persist in the population.

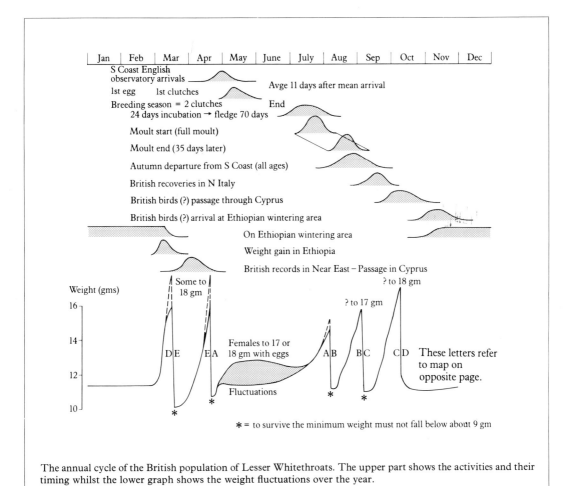

The annual cycle of the British population of Lesser Whitethroats. The upper part shows the activities and their timing whilst the lower graph shows the weight fluctuations over the year.

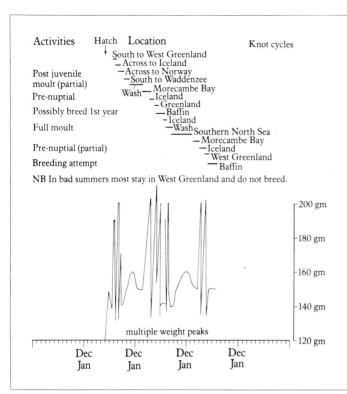

Activities   Hatch Location   Knot cycles

             South to West Greenland
             — Across to Iceland
Post juvenile   — Across to Norway
moult (partial)   — South to Waddenzee
Pre-nuptial   Wash — Morecambe Bay
              — Iceland
Possibly breed 1st year — Greenland
             — Baffin
Full moult   — Iceland
            — Wash Southern North Sea
Pre-nuptial (partial)   — Morecambe Bay
             — Iceland
Breeding attempt   — West Greenland
             — Baffin

NB In bad summers most stay in West Greenland and do not breed.

multiple weight peaks

200 gm
180 gm
160 gm
140 gm
120 gm

Dec   Dec   Dec   Dec
Jan   Jan   Jan   Jan

The Knot, a medium sized wader with a complicated migration pattern, shows multiple weight peaks corresponding to different steps on migration and a winter 'insurance' peak when fat is accumulated.

Ducks do not seem to do this but they are quite likely to pair on the wintering area and then to migrate to the breeding area together. As an example of the annual cycle of waterfowl, a diagram showing the Canada Goose, a familiar migrant over much of North America, has been chosen (see page 105).

It is not possible to describe the life of many seabirds in terms of an annual cycle since the youngsters do not start to breed and undertake a regular annual timetable for several years. The chicks of many species are looked after by their parents long after they have fledged and family parties may undertake their migration together. This period of parental care is of great importance as the young birds are learning the skills of fishing in different areas and under different weather conditions. Ornithologists who have made detailed observations on terns in Africa as late as March – eight months after the young birds hatched – have shown that the adult birds are still much more likely to succeed when they dive for a fish than the young ones. At the time of the first breeding season after

they were hatched some young birds may still be on the wintering grounds thousands of miles south of the natal area. The next year they may return to the breeding area for a short period at the end of the season. In subsequent years, these visits will become earlier and earlier until, after possibly two or three years, the young birds may start to defend burrows and take an interest in pairing. Even then the young pair may return for two or three years before starting to lay and attempt to raise young. The pre-breeding period is not quite so protracted in most species and some terns may breed in their second summer at an age of just under two years.

The cycles, over a six year period, of a young Gannet taken from the British breeding population is illustrated (see page 157), as most is known about this seabird. A similar situation exists in the rest of the northern hemisphere for related species. A most interesting first migration is carried out by the young Guillemot. As with a number of the other alcids, both in the Atlantic and

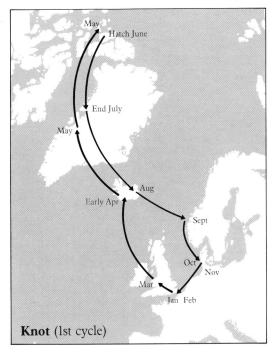

Knot (1st cycle)

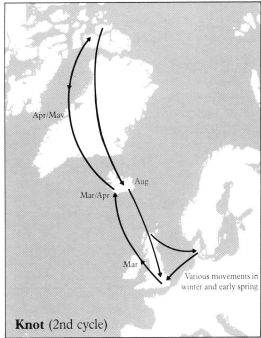

Knot (2nd cycle)

in the Pacific, the breeding birds nest in areas where there is a good food supply during the early part of the breeding season but, since the food is mobile (fish), later in the summer the area of abundance has often moved. The response of species like the shearwaters and petrels is to have the adults undertake long feeding trips bringing food from hundreds or sometimes more than a thousand miles away.

The auks have a different solution. The chicks of many species develop their wing-coverts and body plumage quickly and jump, totally incapable of real flight, into the water to start their migration swimming with their parents. Some of the smaller Pacific alcids enter the sea a few days after they hatch but the Guillemots wait on their cliff ledges until they are about three weeks old. At this stage they have been supplied with a large amount of food by their parents and, therefore, have a substantial fat reserve. However, they are still only half developed. Their ungainly drop/glide into the sea is usually accomplished at night to thwart the attention of the gulls which would undoubtedly prey on them

Knot
This medium-sized shorebird is a regular migrant from the high arctic of Canada across to the western coast of Europe during the autumn, winter and early spring. The maps show the hypothetical cycle of, on the left, a young bird from hatching to its first return to the breeding grounds – probably too late to breed. On the right is the same chart for an adult which arrives on the breeding grounds early enough to rear young. For the population to remain viable it is not sufficient for the arctic wilderness to be preserved but also the transit areas – in western Greenland, Iceland, the Netherlands and Germany and in Britain. Other populations move along different routes – for example the Siberian birds may pass through western Europe on their way to winter in West Africa.

in daytime. It is very important that they should remain in contact with their parents, usually only one of them, for they will swim together for several weeks whilst the young-ster grows its first flight feathers and the adult undergoes its complete moult. Many Scottish birds actually swim across the North Sea to southern Norway at this time. By the winter both the young Guillemot and the young Gannet will have left their parental

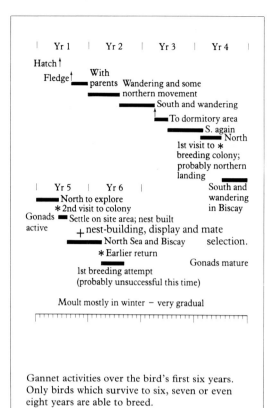

| Yr 1 | Yr 2 | Yr 3 | Yr 4 |

Hatch↑
Fledge↑ — With parents
— Wandering and some northern movement
— South and wandering
— To dormitory area
— S. again
— North

1st visit to ✱ breeding colony; probably northern landing

| Yr 5 | Yr 6 |

— South and wandering in Biscay
— North to explore
✱2nd visit to colony
Gonads — Settle on site area; nest built
active + nest-building, display and mate selection.
— North Sea and Biscay
✱ Earlier return
Gonads mature
— 1st breeding attempt
(probably unsuccessful this time)

Moult mostly in winter – very gradual

Gannet activities over the bird's first six years. Only birds which survive to six, seven or even eight years are able to breed.

Gannets over the colony at Bass Rock in the Firth of Forth, Scotland. These birds live for many years and most will only ever come to land at their own breeding colony. This puts a great premium on accurate navigation.

care and be well on their way to the furthest distances they ever reach away from the breeding area. Successive winters are likely to see them closer and closer to the breeding grounds as the earliest birds to arrive in breeding condition at the colonies are the most likely to be able to secure the best breeding sites.

Individual variation

There are very few biological events which are precisely timed and so it is with the annual cycles of birds. This means that at any particular time of the year some birds of the same population may be as much as five or six weeks behind others. This cannot be anything but an advantage for the population as a whole for, if all the individuals were to be precisely synchronized it is very likely that all might encounter a particular hazard together. If this were simply a severe period of rain which washed out many nests, it might quickly be made good. However, if it was a sudden storm during which the migrating birds perished, the whole of the local population might succumb at the same time. Taken with the variations shown between southern and northern populations of the same species, this means that there may be a very prolonged passage period for a species passing through the southern part of its breeding range or, in some cases, even as it leaves its winter quarters. In Britain a south coast bird observatory might record its first migrant Willow Warbler of the spring at the end of March but still have birds destined to breed in Scandinavia passing through at the end of May, whilst local birds were feeding newly hatched young. The autumn movement of local youngsters would be in full swing by mid-July, with peak numbers of birds migrating in August, but eastern vagrants, of the same species, might still occur into late October. Thus the single species has a passage period through the same site of almost half the year.

This extended spread is most obvious in widely distributed species with large populations. Certainly there are many examples of rarer species which are often found regularly, accurate to within a few days, at the same sites

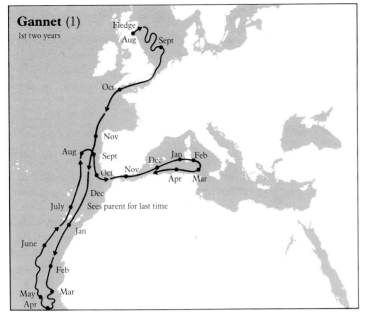

Gannet (1)
1st two years

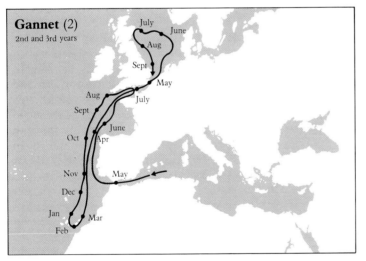

Gannet (2)
2nd and 3rd years

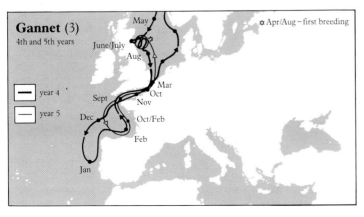

Gannet (3)
4th and 5th years

year 4
year 5

✿ Apr/Aug – first breeding

Gannet: the lifetime track
These three maps show the hypothetical lifetime track of a Gannet
bred on Bass Rock from its fledging
to its first successful breeding, back
on Bass Rock, six summers later.
1) Gannets only raise a single chick at
a time, and having fledged, the young
bird moves about the North Sea and
down the west coast of Europe. In
November or December it sees the
last of its parents who, until then
have been teaching it their feeding
skills. The young bird continues
southwards and, even during the
following summer may still be off
northwest Africa. Much of its first
winter happens to be spent in the
Mediterranean.

2) During its second summer it
ventures up the Bay of Biscay and
into the English Channel. Some may
penetrate the North Sea and even
come within sight of their natal
colony. Its winter movements take it
only down as far as southern
Morrocco and, during its third
summer it reaches the North Sea
and comes to land, for the first time,
late in the season on Bass Rock.

3) The next winter, its fourth, is
mainly spent off North Spain and by
March it is back in the North Sea
reaching Bass Rock in May. Here it
starts to defend a nesting territory on
the edge of the main colony and may
well find a mate. That winter, its
fifth, is spent in the Bay of Biscay
and in early April it is back on Bass
Rock to make its first breeding
attempt – a fully adult black and
white bird with a faintly golden
head and nape.

157

year after year. Birds which breed in high latitudes have such a short period when the weather is good enough for breeding that they have to be very highly synchronized. If the weather proves to be particularly bad, they may fail to breed at all for a year. This may also happen to seabirds if food supplies fail and to predators, like skuas and owls, if their prey species decrease heavily in numbers. Even in temperate areas common species may have to modify their annual cycle to account for the vagaries of the weather. Severe drought conditions can stop hirundines from breeding becáuse there is no mud available for nest building. Drought may also force various species which normally feed on earthworms to abandon breeding as the worms dig deeper into the ground seeking moisture. It may even allow seed-eating birds to breed earlier as seeds ripen faster.

These variations may be induced by many different factors. Individual birds may be genetically inclined to migrate earlier, breed earlier or moult quicker than others. Some may respond slightly differently to the variations in the environment which have their part to play in the triggering of physiological changes within the bird. The birds' own previous experiences will also influence its timing in several ways. For instance a bird whose initial nesting attempt failed and which then lays a repeat clutch will necessarily lag behind those whose first attempt was successful. If, through its initial failure, it were not to have a second clutch after the repeat nest succeeded it would then start moulting earlier than other birds who had a second clutch. An older bird may be able to speed up its preliminaries to breeding and so complete its breeding attempts early. Finally, a really healthy bird will probably always be able to complete stages of its physiological cycle better and faster than one handicapped by an infection or trauma.

Individual variation, as demonstrated by Gwinner with his Chiffchaffs, also applies to the birds' destination, even with siblings, on their winter migration. It equally applies among local populations of birds with respect to their winter quarters. If one could map with a complete one to one correspondence the summer and winter sites of a species across the whole of its range you would get some surprising results. There are, however, plenty of examples, taken from ringing recoveries, to show a simple sort of relationship – birds from the western part of the summer range are more likely to be found in the western part of the winter range. The map of Osprey movements from Europe to Africa illustrates this clearly (see page xxx). It is really only amongst the species which migrate as family parties and those that are gregarious diurnal migrants that any sort of precise tradition can build up. Accordingly, it is not surprising that the geese and swans are amongst the most precise migrants, often coming to the same pond or field year after year when there are other apparently suitable sites close-by. The traditional gatherings at roosting sites of hirundines are rather different. They almost certainly depend on the complex aerial displays which the roosting birds use to advertise the reed-bed roost sites. Certainly ringing has proved that the same sites are used year after year by birds from the same areas – often hundreds of kilometres away. However, this may be coincidence for the birds concentrate into relatively few roosting sites whilst on the move.

Turning the annual cycle

Let us examine, starting with the arrival of the migrant in its summer quarters, the advantages and disadvantages of various alterations to the annual cycle. Some variation in timing will be present in any natural population and changes may gradually evolve as the species encounters different conditions – either through pioneering new areas or through local changes in conditions.

+ Advantages – Disadvantages

Arrive earlier to breed
+ Obtain the best breeding territories and have a longer time to breed.
– Be at risk of starvation if the weather is cold and thus, possibly, be in a worse condition by the time later arrivals return.

Breed as quickly as possible
+ Possibly fit in a further or repeat breeding attempt if the first fails.
– Female has less time to produce eggs and

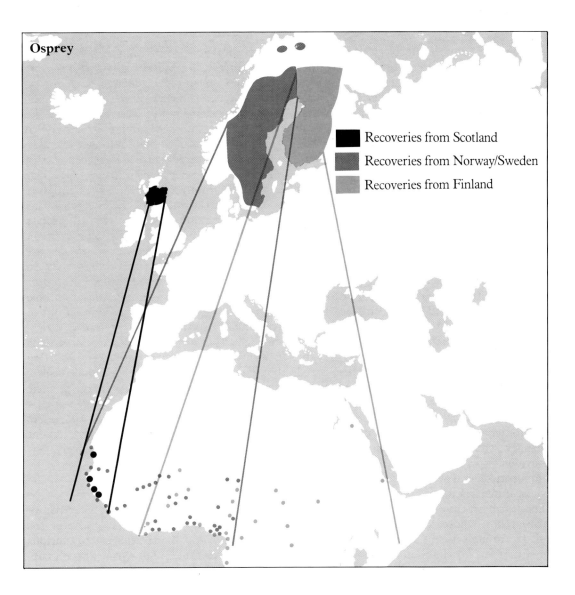

Osprey

Recoveries from Scotland

Recoveries from Norway/Sweden

Recoveries from Finland

thus eggs may be of poorer quality; more chance of disruption to breeding attempt through late cold weather.

Curtail period of parental care after hatching
+ Allow adults further breeding attempt.
− Possible failure to pass on skills, or sufficient reserves of food, to offspring which will already have had at least 35 days of care by both parents − egg and chick − invested in them.

Shorten period of adult's complete post-breeding moult
+ Possibly postpone main moult until win-

Osprey winter areas
The Osprey is a summer visitor to northern Europe. In Scotland it was extinct for almost fifty years but has become re-established thanks to the efforts of conservationists. The wintering areas of three populations have been established from ringing recoveries. The birds from Scotland in the west, Sweden in the middle and Finland in the east have been found across sub-Saharan Africa bearing the same sort of relationship to each other as their breeding areas do. This sort of east to west mapping is not surprising for a species with a broadfront migration pattern and has been demonstrated for some other species. Even with birds making their migration along defined routes, like the White Stork, there is good evidence for clustering of birds from particular breeding areas into the same regions in the winter.

ter quarters reached, thus allowing later breeding or earlier migration in the autumn.

- Autumn migration may have to be undertaken on old feathers. The correct food needed to supply sulphur based amino-acids, for synthesizing keratin (the material from which feathers are made) may be in short supply in winter quarters but abundant in the summering area in the autumn. Fast complete moult may impair flying ability and make adults prone to predators.

Migrate earlier

+ Avoid unseasonably early cold weather.
- Unable to utilize available food resources on breeding grounds late in the season and even arriving too early for stop-over food resources on the journey southwards.

Migrate in shorter steps to avoid having to put on so much weight

+ Less time to put on weight, shorter migration sectors.
- There may only be a restricted suitable area for stop-over; fat reserves may be important *after* arrival at first destination.

Migrate later

+ Autumn insect food resources often seem to outlast the migrants – a resource unused which would allow for a less hectic breeding season.
- Food resource not reliably available; if cold weather comes late birds are doomed and, in any case, may not be able to utilize peak of food resources later on in their migration.

Moult in the winter quarters

+ Much longer time available for moult than during the summer.
- Many wintering birds remain very mobile – this would be hampered by moult. Food resources not as good as during the summer.

Survival through the winter

The whole object of migration is to enhance the survival of the species which performs the migration. We have seen many examples where there is absolutely no chance of the birds concerned being able to spend the winter in the areas where they have chosen to breed. This may be because of snow or ice cover in the northernmost areas or because of the lack of insects and other suitable food in more temperate parts. Seabirds may depend on particular food resources which themselves move in the oceans either of their own accord or as they are drifted by the currents. In some cases, the weather conditions and sea temperature alter to provide the appropriate stage of development in the prey.

The survival of the migrant species in extreme cases is thus simply not possible without migration or a drastic change in breeding area to ensure that food is available throughout the year. The species which breed in high latitudes and migrate southwards to winter in tropical or sub-tropical areas may seem to have a 'soft option': stay in the wintering area and breed there. That this is not the case has been shown for a wide variety of species in both the Old and New Worlds. The problem may simply be expressed as 'Where would the newcomers fit in?' Tropical and sub-tropical areas are richly endowed with their own local breeding species which are, for the most part, sedentary. These local birds are able to set up and defend territories in their home areas and to specialize on particular habitats and food resources. Indeed competition is very fierce: there may be dozens of similar species breeding within a small area able to survive because of specialization within the group on particular ecological niches. One example is the situation met by European Swallows which winter in South Africa. Those that migrate to the Cape area may leave a part of Europe where the only relatives seen are Sand and House Martins but winter in an area with eight other species of swallows and martins breeding.

This immediately poses another, very different, question. If the migrants are arriving to winter in an area with so many local species how are they able to fit in? Some figures show just what a huge problem this entails. Reg Moreau spent the last years of his life gathering together as much information as he could find about the system of migration between the Palearctic and sub-Saharan Africa. His best estimate for the number of migrants reaching Africa during the autumn was a staggering 3750 million individuals. They had about 20 million square kilometres to live in for the winter, which works out at two birds per hectare. Each migrant, therefore, has, on average, a space of about 100 yards by 50 yards to live in. Moreau's figures related only to passerine and other land-bird migrants and allowed for mortality between leaving the breeding grounds and arriving in Africa. His common passerine estimates were as follows:

Willow Warbler	900 million
Sand Martin	375 million
Tree Pipit	260 million
Spotted Flycatcher	250 million
Swallow	220 million
Blackcap	200 million
Garden Warbler	200 million
Lesser Whitethroat	150 million
Common Whitethroat	120 million
Redstart	120 million
Wheatear	120 million
Ortolan Bunting	120 million
House Martin	90 million
Yellow Wagtail	70 million
Whinchat	45 million
Pied Flycatcher	30 million

He allowed a further 1000 million for the extra 50 species of passerines for which he was not able to make an estimate and another 200 million for the near-passerines, the majority of them probably being Swifts. The raptors, mainly the small falcons, added a further 40 million or so to make a total departure each autumn of some 5000 million birds. This did not take into account water-birds such as ducks, waders, gulls, terns, and storks.

Of course, the 20 million square kilometres of Africa are very varied, ranging from the heat and vigour of the equatorial forest to the chill mountain peaks of Kenya. Except for the effects of altitude the main factor which affects the birds' use of the different areas is the amount of rainfall and its timing. The map (see page 162) shows that most areas have a single rainy season: north of the equator it occurs during the northern summer whilst south of the equator it occurs during the northern winter. Twin rains happen along the coast of the Gulf of Guinea and through the Congo basin, most of which is covered with wet lowland forest. The other twin rain area is in central east Africa from northern Tanzania into Somalia where the major rainy season is around May and a minor one also occurs in October or November. The rains mapped are based on a level of 50 millimetres or more in a month – this is not necessarily very much and means that the area may be very, very dry outside these months. Temperatures may be very high too. Indeed the extreme that Reg Moreau featured was a White Wagtail breeding in Iceland with a July mean temperature of about 9°C which may winter in Senegal (where ringed birds from Iceland have been found) with an October mean temperature of 21°–30°C warmer than the breeding grounds. In fact, the mean increase from breeding area to wintering area of 50 species of migrants is over 6°C. The rains and temperatures over different parts of the continent have, of course, formed a range of different vegetation belts which will, themselves, influence the areas where migrants are able to live (the main areas are shown on the map on page 162).

The majority of the migrants arriving south of the Sahara are insectivorous with feeding techniques used on the breeding grounds which are equally applicable in their wintering area. There are some obvious temporary flushes of food which the migrants are able to cash in on in a big way. The plagues of locusts are often mentioned in literature but their availability is very patchy indeed; there may be periods of years in some areas without any appreciable build-up in numbers. Undoubtedly of much greater importance are termites and ants which are generally available everywhere after the rains when they swarm. Other flushes of insects near water (temporary or permanent) can also be available to flocking migrants. These foods are also available to the local birds but, since they are temporary, they cannot be exploited to the full by resident species. Migrants may thus be able to exploit them over a wide area as they become available.

Rainfall (quantity and season) in Africa south of the Sahara is very important in determining which areas are attractive to migrants from the north.

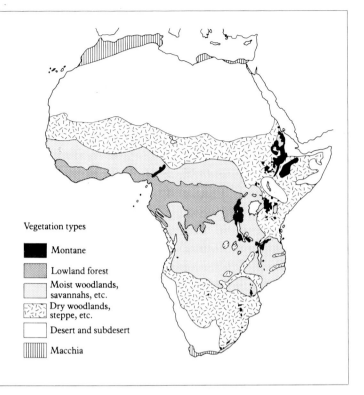

Vegetation types

▮ Montane

▦ Lowland forest

Moist woodlands, savannahs, etc.

Dry woodlands, steppe, etc.

☐ Desert and subdesert

▥ Macchia

The vegetation zones in Africa are partly determined by rainfall but also depend on other factors – like altitude and soil type. Their effect on the distribution of migrants is crucial.

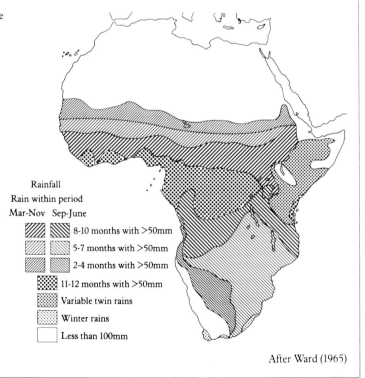

Rainfall

Rain within period

Mar-Nov Sep-June

◫◫◫ 8-10 months with >50mm

◫◫◫ 5-7 months with >50mm

◫◫◫ 2-4 months with >50mm

▨ 11-12 months with >50mm

▨ Variable twin rains

▨ Winter rains

☐ Less than 100mm

After Ward (1965)

However, there are in fact many species of migrants that do not travel long distances seeking food. They tend to take up a territory or to stay in a particular area. Here they may compete for food with resident species although few cases of overt competition have been reported. Perhaps the major problem, named 'Moreau's Paradox', is that many species of migrants reach west Africa shortly after the summer rains with their reception area holding plenty of food. As their stay progresses the food supply gets steadily less and less and yet, at the end of their stay with the local habitat apparently in a very poor state, these migrants are not only able to support themselves but also put on considerable amounts of pre-migratory fat for their crucial journey northwards. For some species at least part of the answer is in their ability to exploit fruit from such scrub shrubs as *Salvadora persica*, *Maerua crassifolia* and *Balanites aegyptiaca*.

Many ornithologists in Africa have remarked on the preference of northern migrants for 'edge' areas where human activities have modified the ecology of an area. The history of human activity over the continent is relatively short in its most extensive modifications of the ecology but, during this short period, Europe has itself changed greatly. Both changes should, in general, have allowed more and more long-distance migrants to exist both in summer and in winter. Certainly when the primeval forests covered temperate Europe there would have been many fewer long-distance migrants; now there are so many more areas for scrubland species to breed (not only in scrub but in hedges too) and buildings for Swallows, Swifts and House Martins to use. Of course, over the last 2000 years, the Sahara has gradually extended southwards and several million square kilometres of suitable wintering area have been lost. The last manifestation of this, the Sahelian drought of the 1960s, has had a very definite effect on the numbers of several species which wintered in the Sahel. Species like the Common Whitethroat which was known to use the berries of *Salvadora* for fat deposition in the spring crashed over much of western Europe. What

seems to have happened is that the bushes were able to survive several years of drought but then died – or were used as fodder for dying stock or as firewood by the desperate local human population. Without the bushes the birds were not able to make the necessary fat level to be able to cross the Sahara. Such disasters, causing the British population to fall by 75 per cent over one winter, are uncommon and not easily reversed. Whitethroat populations have only recovered slightly since the Sahelian ecology cannot recover quickly – mature bushes can only be replaced over a period of several decades.

The situation facing the migrants from North America to the tropics is even more complicated than that in the Old World. Tropical America has a very much more complicated system of relief and thus a much more varied ecological spectrum available for the birds. This is reflected in the immense diversity of species breeding – the richest area in the world. Detailed investigations into migrant *versus* resident bird interactions have blossomed over the last ten years and it has become apparent that many of the migrants from further north really have to be considered as part of the southern avifauna which happens to move northwards to breed. In several studies of areas close to the southern United States, for instance in the West Indies and in Mexico, the impact of the migrant birds has been shown to have a very strong effect on the local breeding birds. Many of the wintering migrants are 'small foliage gleaners' – small insectivorous birds which take their food from the leaves of trees and shrubs – and this type of bird is very much under-represented in the local breeding population. This very clear impact gradually becomes less and less marked the further south one goes. By the time that the central northern parts of South America are reached migrants form only a small part of the local avifauna. The restricted availability of land for migrants not reaching the main part of the South American continent has its part to play, as may clearly be seen on the map on page 164.

In Amazonia the resident species include very large numbers of specialist and generalist fruit-eaters, as well as assemblages of insectivorous species, which often forage together. These guilds of birds are sometimes associated with social insects, particularly ants and termites. Within the guild the different species keep together but are each feeding in a slightly different manner to avoid competition between members. Such guilds may be joined by migrants during the winter and, further north in Central America, migrants may make up a considerable part of some guilds or even join together on the wintering grounds to form their own. Many research workers reporting from Panama southwards have found that the migrant species generally form a minor part of the avifauna at most sites and, where they join guilds, are generally subordinate to the resident species. Furthermore migrants are generally found in disturbed habitats whose ecology has been altered within fairly recent times by man, in the higher areas where the resident birds are generally not so diverse (lowland forest) and in areas which show some seasonality – like arid scrublands subject to short rainy periods. Further efforts to quantify their impact has been attempted by spot counts through the year. From Panama southwards, studies have usually found less than 10 per cent of the birds recorded during the year were migrants with peaks of less than 30 per cent as the migrants arrived in late autumn. Indeed undisturbed lowland forest areas generally showed many fewer than this. Since, on average, the migrants weighed about one third of the average resident species their effect on the bird biomass was much less even than the ratios imply.

This interesting contrast between the northern area, where North American long-distance migrants winter, and the southern area, where different longer-distance migrants are found, is intriguing. To the north, in Mexico and the Caribbean, there are many species which arrive and occupy a position in the local avifauna which is vacant when they leave to breed in northern areas. Detailed research has shown that wintering birds here may often take up territories (both males and females) and defend them or become active in guilds which may or may not include resident species. If they do then the migrants do not seem to be inferior. Further south the percentage of migrants in the avifauna is very much reduced. They appear to be able to winter only by using changing resources and by being mobile; they often use disturbed habitats and, if they join guilds, they tend to be subordinate within them.

Similar problems also face the seabirds when they move to areas offshore in distant parts but, for them, there may be super-abundant food supplies which they and the resident species are able to exploit together. There is also the possibility of the migrant birds staying very far from the shore and out of reach of all but the most far-foraging breeding species. Migrant waders are often

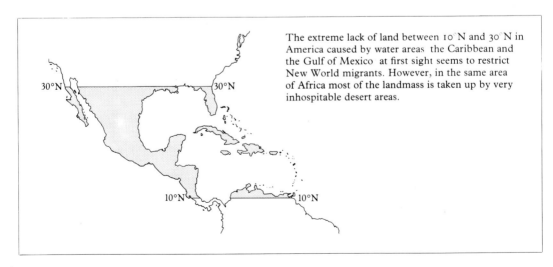

The extreme lack of land between 10°N and 30°N in America caused by water areas the Caribbean and the Gulf of Mexico at first sight seems to restrict New World migrants. However, in the same area of Africa most of the landmass is taken up by very inhospitable desert areas.

in a very good position since they arrive on their wintering grounds to find that there are very few resident species present, certainly not in any numbers. Their morphology is generally such that there seems to be little competition on the estuary or mudflat and it does seem as if they have evolved to reduce such competition. Certainly if they are high arctic breeding species where the food supply may be very abundant there would be no need of a specialized feeding technique.

Survival on migration

The risks involved in the migration flight really start with the period of fat deposition which, for a long-distance migrant, may start several weeks before departure. Although a fat bird may be a very efficient long-distance flying machine it may also be a much less agile item of food for a predator! The migration flight itself, for the young birds literally a flight into the unknown, is obviously full of hazards: bad weather may cause the birds to stray from the correct path; faulty navigation or insufficient fuel reserves may cause birds to settle in the wrong place; and the bird may fail to adapt to the different surroundings it encounters at stop-over points or on the wintering ground.

The small migrants travelling from Europe to the Mediterranean have a further hazard to face. The Eleonora's Falcon is a small falcon breeding on cliffs and rocky islands in the Mediterranean. The whole of its life-cycle is designed for it to breed late in the season so that it can feed its growing young on the autumn migrants passing through the Mediterranean. Although it is a rare and local bird, breeding at less than a hundred colonies, there is a population of about 4500 pairs breeding each year. With their young this means that there are about 20,000 Eleonora's to be fed through the autumn on small migrants. At five to ten small birds a day for each falcon this means that an astonishing five to ten million birds fall victim to them each year. The distinguished German ornithologist, Erwin Stresemann, estimated that more than one in every 600 small birds setting off to migrate across the Mediterranean finishes its life as falcon food.

Such hazards are a real threat to the bird populations which do migrate and yet, unless migration only involved 'acceptable' risks, it would quickly cease to be adopted as a sensible option. The fact that a majority of the world's birds are thought to be migrants over at least some part of their range shows that it is still a successful option. Obviously its most potent attribute is that it allows birds to breed in areas where they are unable to survive outside the breeding season but which provide good conditions, and often even a super-abundant food supply, for a relatively short period.

Perhaps the example of the Stonechat and Whinchat populations breeding in Britain are the best proof of just how successful migrants can be at surviving. Of these two birds, very much the same in habits and habitat, the Stonechat is a resident or short-distance migrant staying in Britain or adjacent parts of western Europe whilst the Whinchat is a full scale long-distance migrant crossing the Sahara to spend its winter in west Africa. Detailed studies show that the Whinchats are able to survive twice as well through the winter as the Stonechats, which have to breed more frequently and more successfully than the Whinchats to survive as a species.

Conclusion

Migrant birds have the whole of their annual cycle modified by the needs of migration. Their breeding season is often shortened, as compared to non-migrant or short-distance migrant species. They mostly have to adapt their physiology, in the weeks before migration, to ensure that they store the fat needed as fuel on the journey. At the very least they must be able to live in two widely separated areas – their summer and winter quarters. Indeed many species have several stop-over points between their breeding area and their extreme southern penetration at mid-winter. In these places they may find weather conditions and living areas very different from those experienced during the summer and they must be able to find their food and survive. Thus migration can truly be said to form the migrant's annual cycle rather than forming a part of it.

The evolution of migration

The complex patterns of bird migration that exist in the world today have evolved as the birds concerned have themselves developed. The time-scales involved range from the year to year changes, which may be imposed by weather patterns, to the millions of years that the continents have been moving on the face of the earth. However, the fascinating influences on the birds' migration patterns are but a part of the story, for the birds' themselves have also evolved to be able to perform their migrations successfully. This has involved changes in physical appearance (for example wing shape), in physiology (for instance pre-migratory fattening), and even in genetically controlled rhythms (like the *zughenruhe*).

The start of migrations
Reg Moreau, in his classic book, *The Palearctic-African Bird Migration Systems*, notes that any writer on bird migration is always likely to be asked the question 'How did bird migration begin?'. This may seem unanswerable but it is not. The simple answer is that nature abhors a vacuum; if there were no birds in higher latitudes during the summer to exploit the flush of insect food there would be a vacuum on a huge scale.

Indeed the existence of seasonality must, very quickly, produce migrant populations, of any species, to exploit the seasonal food resource because they are unable to live in the area where that resource is only available for part of the year. A very simple model, shown in the diagram (opposite page), actually predicts what has, to many people, been a puzzling aspect of migration: the leap-frog effect of the northernmost breeding populations of a species actually wintering further south than any others. As seasonality in the environment developed, a northern area became suitable for the species to breed. The birds that colonized it were forced out as the conditions deteriorated in the autumn and moved southwards. When they reached more clement regions local breeding birds of the same species, which were already there, were able to claim the available resources forcing the northern breeding birds further south. A simple extension of this argument also accounts for the start of altitudinal migration.

Seasonality is an essential ingredient of this 'start of migration' model. In temperate regions the summer/winter cycle of temperature is its most obvious aspect but seasonal rains, the seasonal flowering or fruiting of plants or even the fluctuations in fish spawning could all be enough to trigger migration in a species of bird. Once any system of migration has been initiated, if it is to be a successful strategy, the birds which are best able to practice it will be able to survive better than those that are inefficient. The migration system and the birds, therefore, will develop and adapt. Migration has developed not only in birds but in mammals, fish, insects, crustacea – in fact in almost every form of animal studied – and it seems likely that migratory birds have lived for tens of millions of years. This is a very long time for all sorts of complicated patterns to emerge although the tremendous changes of climate over the last few thousand years will inevitably have an over-riding influence.

The history of the bird species over the temperate regions of the northern hemisphere can be traced back to their evolution over 150 million years ago when the earth itself was undergoing tremendous changes. Birds first started to fly after the massed continents, Pangaea, had broken apart and started to drift towards their present positions (see the maps on page 168. In the Cretaceous period 100 million years ago, the map shows that when the ancestors of many present-day species were evolving, the land mass was different from today. In fact relative posi-

tions of the Palearctic-African have remained very much the same over the whole period but, 100 million years ago, North and South America were far apart. At this time India broke away from its position off south-eastern Africa and started to drift towards Asia. During the Eocene, 50 million years ago, the two parts of America were beginning to come together and India was half-way to Asia. Over the last few million years the Continents have continued to move. The collision between India and Asia produced the massive mountains of the Himalaya. Australia moving away from Antarctica had hardly any effect on the majority of Holarctic birds but the continuing expansion of the Atlantic will certainly have been very important for seabirds and may have eventually stopped regular migrations between West Africa and eastern South America.

Whilst all this was happening there were staggering changes in climate, both on a global and local scale, whilst the zones of vegetation also changed dramatically. These will have had a tremendous effect on the local avifaunas but it is still possible to see how the North American avifauna owes more to the Old World than to South America. In a classic analysis of the origins of the birds breeding in North America, Ernst Mayr paid particu-

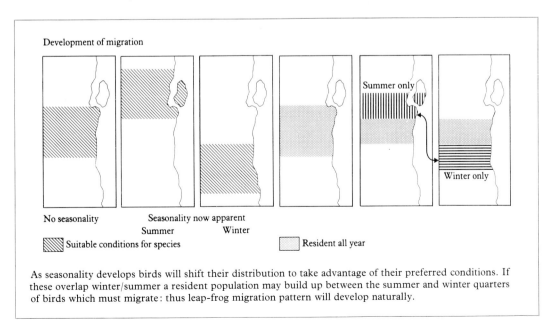

Development of migration

No seasonality

Seasonality now apparent
Summer   Winter

Summer only

Winter only

Suitable conditions for species   Resident all year

As seasonality develops birds will shift their distribution to take advantage of their preferred conditions. If these overlap winter/summer a resident population may build up between the summer and winter quarters of birds which must migrate: thus leap-frog migration pattern will develop naturally.

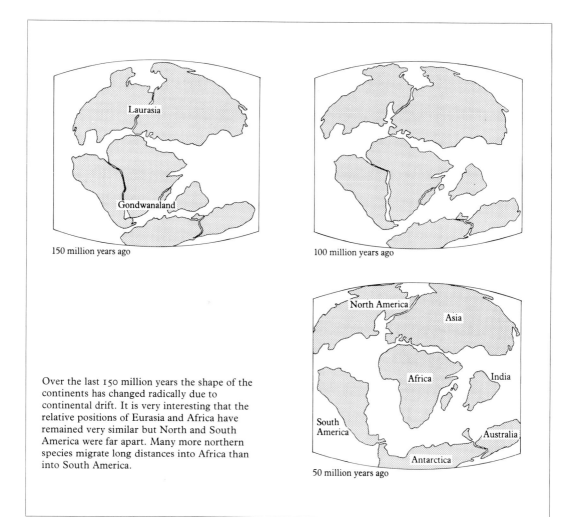

150 million years ago

100 million years ago

Over the last 150 million years the shape of the continents has changed radically due to continental drift. It is very interesting that the relative positions of Eurasia and Africa have remained very similar but North and South America were far apart. Many more northern species migrate long distances into Africa than into South America.

50 million years ago

lar attention to the passerines. He took six areas, from Alaska south to Mexico, and showed that only at the southernmost site did species of South American origin outnumber those of Old World origin. He extended his analysis to real numbers of breeding birds and found that the importance of the Old World element was diminished. However, he also showed that this was essentially due to the Old World birds often being rather large, with resident species occurring at low densities, whilst the South American ones were smaller migrants which were found at higher densities. These differences are real indications of the extremely long-term factors which may come into play when we try to consider the origin of bird distributions.

It is not possible to be absolutely certain about the detailed aspects of bird distribution millions of years ago nor, in many cases, to be definite that the species involved were exactly the same as those we see today. Detailed studies of bird fossils, even when only a few shattered bones exist, have enabled paleontologists to be certain that many Eocene species (about 50 million years ago) would be recognizable today, at least to their general group (e.g. pelicans, ducks, herons, owls, waders, cranes etc.). The same sort of studies have proved that many birds alive two or three million years ago were of species which are still extant today. Unfortunately really good bird fossils are only available where there have been particularly favourable circumstances for their preservation. In most cases they will quickly have disappeared for ever since they are much less massive than the almost solid bones of terrestrial animals like the large saurians or early mammals.

However, since we can see that bird populations are of great antiquity, it is instructive to look again at the maps (previous page) of the drifting continents and to see whether there are any aspects of present migration patterns which might fit in with their movements. One of the strangest patterns of migration is the almost complete lack of land-bird migrants from Asia reaching Australia. The maps show that the convenient stepping stones provided by Indonesia have only been conveniently placed in the fairly recent geological past. Similarly the relatively small extent of movements between North and South America, as compared with Eurasia and Africa, may be explained by the great distance between them during the early part of birds' evolution and, later, to the sea gaps in Central America which persisted until about 10 million years ago. This would have isolated the South American avifauna from northern migrants and allowed a much greater diversity of birds to evolve than in Africa, where migrants would have been able to reach the whole continent throughout the period. It is tempting to explain the relatively low numbers of migrants reaching India

because of its rather recent attachment to Asia but this may be only a coincidence. Many species that seem to have started as migrants to Africa have developed healthy populations reaching India. In any case, for many species, the mountains to the north of India may effectively seal the sub-continent off. To the north there has, throughout the period, been a very obvious route for shore birds from Arctic Canada and Greenland to the European and North African coasts but recent glaciations will certainly have extinguished any ancestral bird populations that could have been using this route, especially over the time-scale that the continents have been moving in.

Continental drift may have its part to play in the overall determination of some of the ancient patterns of migration but only after the birds concerned had already been undertaking migrations in any case. With movements of a few centimetres each century, the changes of the earth's crust could certainly not have been perceived by the birds nor would it have had any effect, except over millions of generations. It is climatic changes that will have shaped the movements of each population. Whole populations of birds will have moved back and forth as the conditions changed. However, others will have evolved into different forms to be able to exploit new conditions as they arose and some, inefficient

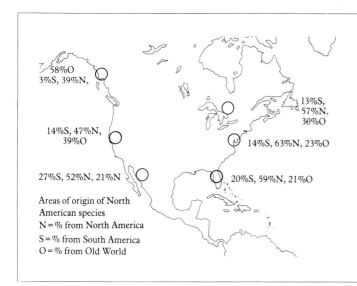

58%O
3%S, 39%N,

14%S, 47%N,
39%O

27%S, 52%N, 21%N

13%S,
57%N,
30%O

14%S, 63%N, 23%O

20%S, 59%N, 21%O

Areas of origin of North
American species
N = % from North America
S = % from South America
O = % from Old World

The overall areas of origin of the birds breeding at six points in North America reveals that species of Old World origin outnumber South American ones at all sites except in northern Mexico – even here more than a fifth are from the Old World ancestors.

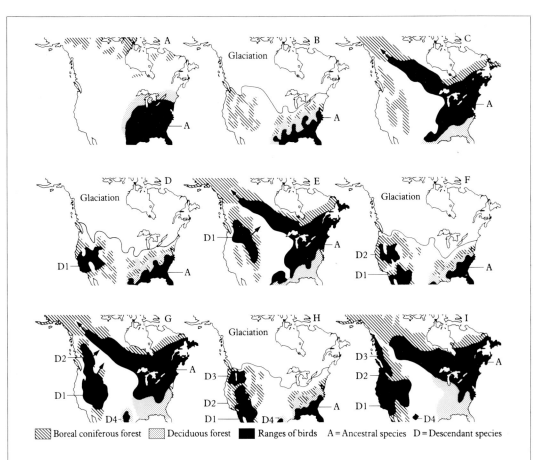

Boreal coniferous forest   Deciduous forest   ■ Ranges of birds   A = Ancestral species   D = Descendant species

A computer simulation of the glacial ebb and flow over North America has been used to predict the changes in distribution of a wood warbler species as the suitable areas of forest have fluctuated.

The ancestral group (A), in the south-east (top left) gradually gives rise to extra species in the west until, after four cycles of glaciation, this ancestral group has three representatives in the west and one (with a very restricted distribution) in the central south. Below the real present distribution of two groups of warblers is shown – both are very close to the simulations!

1 Black-throated Green Warbler	6 Eastern Nashville Warbler	After Mengel
2 Townsend's Warbler	7 Western Nashville Warbler	
3 Hermit Warbler	8 Virginia's Warbler	
4 Black-throated Grey Warbler	9 Lucy's Warbler	
5 Golden-cheeked Warbler	10 Colima Warbler	

Left are the Black-throated Green Warbler group and, to the right, the Nashville Warbler group. With the latter the eastern group and the most recent western ones are not yet different species: they are recognised as separate populations at the sub-specific level.

at following changes in their environment will, of course, have died out. In the recent geological past the most potent effect on bird distributions has been caused by the successive ice ages. The fairly regular advance and retreat of the ice sheets over North America and Eurasia during the Pleistocene has had its part to play in the shaping of the landscape over most of the temperate holarctic. It has produced profound changes in habitat over periods of a few thousand years and, in causing bird populations to shift around the globe which have often become isolated from their original genus, it has also given rise to many new species of birds.

In North America the neatest explanation of just how the waxing and waning of the glaciers can cause new species to form has been provided by an explanation of how two groups of warblers arose (see diagram on page 170). The correspondence between the two groups and the model, admittedly geared to fit the species, is remarkable. The birds concerned are of course mainly migrant species and it is easy to see how their migration strategies will belong to the same broad tradition but have been modified, of necessity, as the glaciations not only altered the summer area suitable for them to breed in but also the ecology of their previously suitable wintering grounds. The history of European species may also be looked at in the same way, especially as, for many species, there is much more evidence to be able to piece together the migration routes with more precision than in America. Successive *maxima* and *minima* in the ice covering will have caused speciation amongst such groups as the *Sylvia* warblers but a consideration of the most recent glaciations is most instructive.

As an example the Whitethroat and Lesser Whitethroat are an ideal pair. One species, the Whitethroat, *Sylvia communis*, exists in two main forms: the eastern race *Sulvia c. icterops* which moves round the eastern edge of the Mediterranean and the nominate race *Sylvia c. communis* which moves around the west edge. The Lesser Whitethroat winters in east Africa (see map on page 172) and migrates around the eastern end of the Mediterranean. The two birds are clearly

very closely related but, equally, are certainly different species which breed together without hybridizing over much of Europe. It is easy to explain both the migration strategies adopted at the moment and the means by which the races of Whitethroat and the Lesser Whitethroat became different. If related species in Europe are examined their present distribution and migration patterns may very often be explained in a similar way. Examples include the migratory divide in the Blackcap, the Reed Warbler and the Marsh Warbler, the Pied and Collared Flycatchers, and the Red-backed and Woodchat Shrikes.

Ancestral homes?

The convenient explanation of migration that used to be given was that the migrants were 'spending the winter in their ancestral homelands'. As can be seen from the pressures put on birds by climatic changes it is very difficult to pinpoint an 'ancestral home' for any species since, taken over any reasonably long period, their distribution will have changed and, in many cases, so will the birds.

It is, however, also easy to see how the changes must have evolved gradually as the ecology of the regions changed. The idea of an ancestral home as being the objective of the migrants' flight is thus not far wide of the mark. It is also a very useful concept which serves to emphasize the very traditional aspect of migration. As we shall see patterns can change over a few years but this almost always seems to be effected by swift evolution. Indeed it is very difficult to see how a species could suddenly start a totally new pattern of migration except under very special circumstances.

The first possibility is where parents and offspring move together, as with the geese. Here it would be possible for a goose to pioneer a new area in one year and then, next winter, take its family to the new site and gradually increase the population using it. If the new site were to be 'better' – that is allowing more birds or the same proportion to survive the winter but to return to the breeding area earlier or in better condition – the whole population might shift after surprisingly few generations.

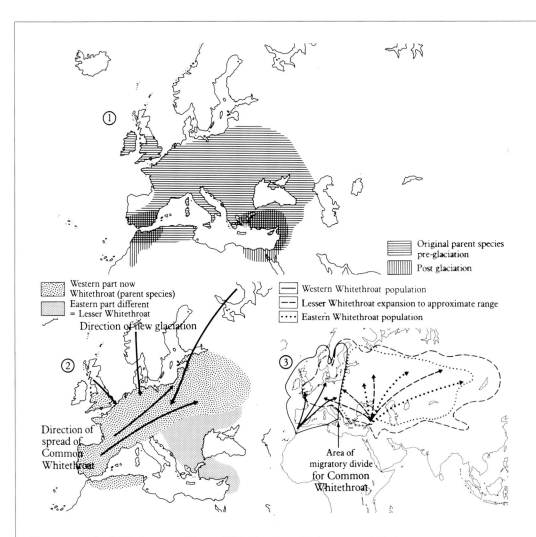

How present-day Whitethroats and Lesser Whitethroats evolved over two glaciations.
1) The original parent species is forced southwards into two distinct populations, at each end of the Mediterranean, by the advance of the ice.
2) As the ice retreats the ancestors of the Common Whitethroat spread from the western refuge well into Asia so that, as the ice returns both they and the Lesser Whitethroat ancestors (already distinct) are forced into the eastern refuge.
3) During the present inter-glacial period western populations of Common Whitethroat (*Sylvia communis communis*) originated from the western refuge and migrate that way. Eastern populations (*S. c. icterops*) migrate through the eastern Mediterranean as do all the Lesser Whitethroats (*Sylvia curruca*).

The other means of establishing an apparently new migration tradition quickly is if a suitable wintering area happens to be found by the pioneering breeders, whether they arrived naturally or were introduced, which happens to correspond with their perception of where their *real* wintering area should be. It would certainly be possible for a migrant species which normally breeds in Britain and winters in Spain to be introduced into North America at a suitable breeding site and thrive. Its winter movements might involve a flight of 900 kilometres SSW accomplished by using magnetic compass and sun compass clues. Such a flight might well take it to a perfectly suitable wintering area in the

United States. This was the sort of result obtained by Perdeck for his inexperienced young Starlings who happily flew to winter in the 'wrong' area although it lay at the correct distance and in the correct direction from the release site to which they had been transported.

Genetic control

We have seen how the majority of migrants are capable of making their migrations as young and inexperienced birds without the benefit of earlier generations to guide them. Indeed even the geese, which do generally remain in family parties, are able to migrate successfully by themselves as juveniles. This implies a high degree of genetic control over all aspects of migration. We have seen experimental evidence of this for migratory restlessness, weight control, moult, etc. in warblers kept under controlled conditions. In warblers and buntings the basis of navigation has been shown to be under genetic control also.

However, for the birds to be able to evolve new strategies, it is really necessary to show that these aspects of their lives are able to adjust genetically in such a fashion that there is a gradual change in behaviour rather than simply succeeding or failing. This Peter Berthold and Ulrich Querner have been able

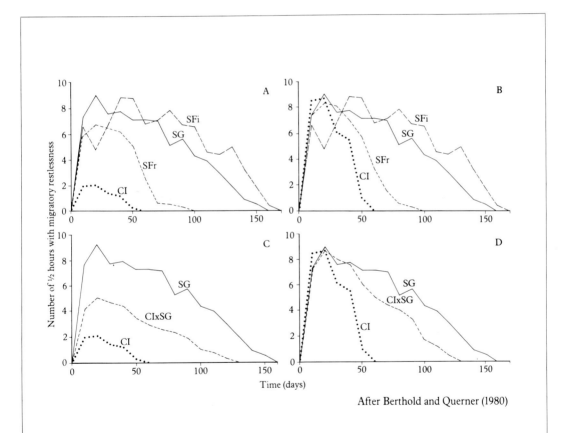

After Berthold and Querner (1980)

This elegant experiment confirms the genetic basis of migration in Blackcaps. In A and C results for all birds tested of each population are added together but in B and D only those birds which showed some restlessness are plotted. Clearly increasing migratory behaviour is shown from the Canary Island to Southern French to Southern German and to Southern Finnish birds. When the German and Canary Island birds were hybridised (CI + SG) their offspring show absolutely intermediate behaviour! SF = Southern Finland; SG = Southern Germany; SFr = Southern France; CI = Canary Islands.

to demonstrate in a very neat series of experiments concerning Blackcaps, from different populations, hand-reared from the nest and kept in aviaries. The important result (see diagram on page 173), that the degree of migratory restlessness from youngsters whose parents come from migratory and non-migratory stock is intermediate between their parents, demonstrates exactly how changes in migratory strategy can be transmitted genetically. It also shows that the genetic basis of partial migrant populations probably lies in a distinct polymorphism rather than in the existence of a continuum between sedentary and fully migratory individuals. This is not an unexpected result since there are often vast areas totally unsuitable for wintering birds on their routes from summering to wintering grounds. A continuum of genetic migratory information might consign part of the population each year to these areas where they would perish.

The effects of competition

So far we have been considering the migrant birds of a particular species by themselves without the pressures they are likely to be subjected to from other species also competing for the resources which they need for survival. During the winter these resources will normally be food supplies although, particularly for birds in high latitudes, safe and warm roosting sites may be also at a premium. During the summer competition may also be centred on the occupation of particular nest sites. Such competition is likely to be both intra- and inter-specific in character. The effects of intra-specific competition, that is the bird competing with members of its own species, includes territoriality which enables the individual birds to partition out the available area between themselves. The equivalent effect of inter-specific competition is generally to reinforce the differences between the two species.

Competition may affect migrants at any stage during their annual cycle. Not only may there be severe competition on the breeding grounds but also, with a different set of birds, in the wintering area. Further they may have to fit in with still more groups of species on

migration – often in different conditions during the autumn and spring movements. The major competitive element comes in securing food. There have been many studies which show just how related and competing species develop different life styles to enable each to maximize its feeding potential. The features measured include preferences for habitat, foraging height (from ground level to tree-tops) and feeding methods. There are many examples of species where the ecological 'niche' occupied is very specialized and in which several similar species compete. However, elsewhere, if only one of the species exists, the single member of the group may be found living in habitats not exactly suited to its specialities.

Alan Keast has recently surveyed the relevant evidence for warblers breeding commonly in Ontario, Canada. His work used evidence from a variety of sources concerning the local breeding birds but there is every reason to suppose that the sort of results he obtained would equally well be applied to warblers, Old or New World, from other areas and, in general terms, to the whole variety of migrant birds. His first analysis concerned the timing of the various parts of the birds' annual cycle. His charts for fourteen species showed that only one of them, the Yellow-rumped Warbler, spent on average more than three and a half months on the breeding grounds. Even this was less than the time it spent in its wintering area, an average of over four months. In many cases the time spent in the wintering area was six or even seven months (American Redstart). In fact the overall averages, for the fourteen species, were just over six months in the wintering area with the rest of the year almost equally divided between the breeding area and migration (see diagram on page 175). Keast also shows that there is relatively little spread in the arrival dates in Ontario compared with the departure dates from Ontario and the wintering area or arrival dates in the wintering area. This is because the summer food supply, for all the species concerned, can only become available after Ontario has started to warm up in the spring. The spread of the mean arrival dates for all fourteen

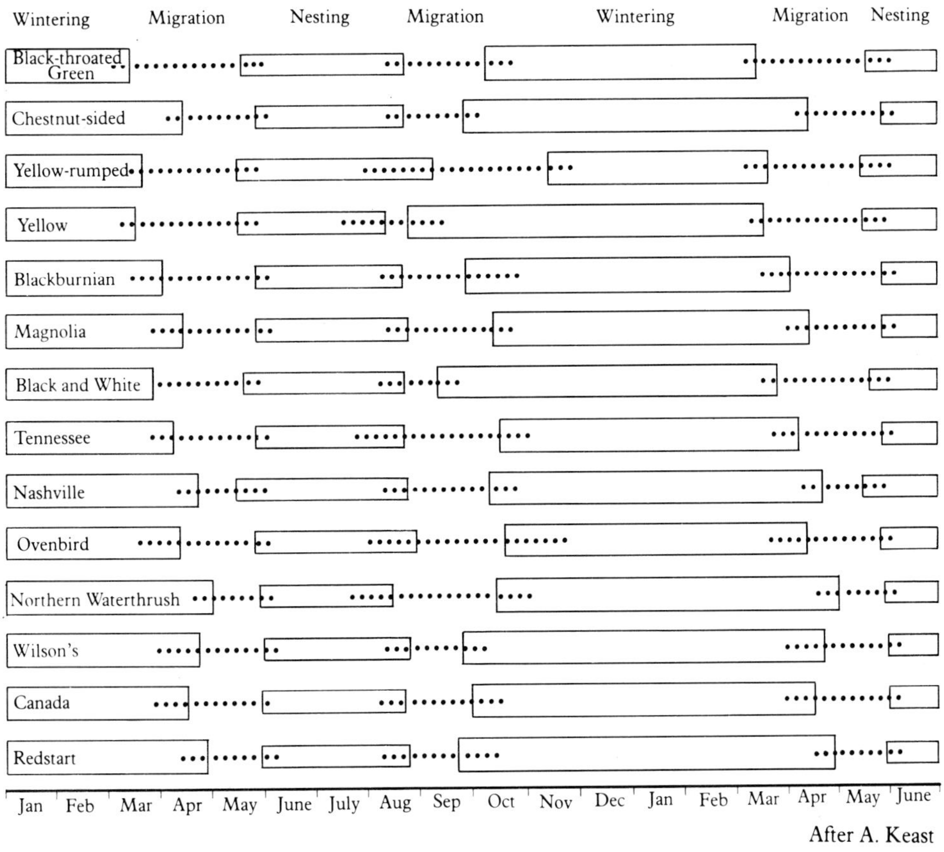
Wintering
Migration
Nesting
Migration
Wintering
Migration
Nesting
Black-throated Green
Chestnut-sided
Yellow-rumped
Yellow
Blackburnian
Magnolia
Black and White
Tennessee
Nashville
Ovenbird
Northern Waterthrush
Wilson's
Canada
Redstart
Jan
Feb
Mar
Apr
May
June
July
Aug
Sep
Oct
Nov
Dec
Jan
Feb
Mar
Apr
May
June
After A. Keast

The comparatively short time spent by these New World warblers in their breeding quarters is clearly shown – less then 100 days on average for many species – most spend very much longer on their wintering grounds – six or even seven months for some.

species is just under two weeks but departure dates straddle four weeks and arrival and departure from the wintering grounds almost 11 weeks and more than six weeks respectively.

Many species occur on migration in Ontario and they are often associated in the loose mixed-species feeding flocks of warblers that may be encountered all over North America during the fall. Detailed observations of the feeding behaviour of these migrants, together at passage sites, provided a measure of the separation from each other that their height and style of foraging provided. Similar detailed studies have, of course, been carried out during the breeding season but these measurements are from birds which only use the area they were measured in for a short migration stop-over (see the diagram above). The fifteen species studied – twelve warblers, two vireos and one chicadee – show a considerable overlap in ecological niche whilst also demonstrating detailed differences from species to species. The particular techniques favoured by any particular species will be controlled by various physical factors. For example, the heavier species tend to feed radially, approaching their prey along branches and twigs, but the very lightest may be able to feed tangentially, perching on the foliage from outside the tree or bush.

Where this ecological approach becomes of particular interest is when surveys at different seasons can be compared to see whether each species maintains its feeding niche throughout the year. Keast was able to obtain data for summer, winter and both migrations for five species and for summer, winter and fall migration for another. The results are shown in the diagram on page ooo, with two different charts for each species during the breeding season and for three species in the winter. These results come from different studies in different areas. For most of the individual species charts shown, the whole range of feeding sites were available but, for Yellow-rumped Warblers at one winter site (the Cays off Belize) and for Yellow Warblers at a breeding site (in Labrador) all the tree sites were much reduced or totally absent. The profiles of the majority of these species show considerable consistency from place to place with only the Yellow-rumped showing a big change – from tree feeding on the breeding grounds to ground-feeding in the wintering area. The three species with the most similar breeding profile in Ontario – Yellow-rumped, Black-throated Green and Blackburnian Warbler – show different fall migration patterns. In the winter, the Black-throated Green Warbler and Blackburnian Warbler are quite alike even though they winter in different areas. However, the Yellow-rumped Warbler, which winters in the same area as the Black-throated Green Warbler, has altered its profile and feeds much closer to the ground.

These detailed findings for New World warblers indicate just how the need to adapt behaviour to the ecology of the various different areas it uses during the year may affect a migrant. Of course the behavioural differences have evolved in the same sort of way that the physical differences between species have arisen.

In the case of the feeding behaviour of these warblers, physical changes in, for example, bill form and shape may also be involved. It is, therefore, not only the effects of competition between the birds themselves that shape their appearance and their attributes.

Physical adaptations for migration

There are some migrants that seem, already, to be superbly adapted for their long journeys. For example, the swifts are totally at home in flight and, since they may fly throughout the day and night whether they are migrating or not, the actual act of migration is a very simple step. There are, however, other species whose life style on the breeding or wintering grounds seems to be totally inappropriate for a long-distance migrant. The shy, skulking behaviour of rails and crakes does not have any bearing on their potential as migrants. When they are flushed from cover, they seem to lack the power to fly quickly or far but rather struggle a short distance to land, in thick cover, with an apparent sense of relief. And yet these same species are often observed at lighthouses at night during migration flying strongly. For this group the power of relatively swift and long-range flight is present all the time but, when flushed from their preferred habitat, be it in winter or summer, it is inappropriate for them to fly fast and far. If they were to they might easily out-fly the bounds of their particular habitat and find themselves over a totally unsuitable area. Their weak and short-distance efforts are, therefore, a behavioural adaptation to keep them safely within a good area.

There are large numbers of birds which have genuinely conflicting requirements between their flights on migration and the style of flight which best suits their feeding ecology. For example, any small bird which needs to be an agile insect feeder making short forays after insects needs to have short and broad wings. The ideal wing shape of the same bird for migrating is long and narrow to give the best flight-range for the consumption of a unit of fat by a bird of the same weight. The effects of this sort of conflict can be seen on the wing/weight relationship for some of the Old World warblers. The sedentary Cetti's and Dartford Warblers have much shorter wings than the Reed and Willow Warblers (both species are about the same weight), which are long-distance migrants. Indeed the detailed drawings of their wings show immediately that, out of a group of four marshland warblers, two are largely sedentary and two migratory. The sedentary species have short, broad wings with a relatively long first primary and the wing point the fourth or fifth feather from the outside (see the diagram below). The two migrants, Savi's and Sedge Warblers, have much smaller first primaries but their longest primaries are the second or third from outside.

The action of these pressures on different populations of the same species can be seen in many birds that have different migration distances over their range. For example, in Britain the Dunnock is a common and widespread bird of hedges, woodlands and gardens. Many thousands are ringed each year and they are very seldom found more than ten kilometres from their ringing site. Nevertheless, during the autumn, British Dunnocks have often been recorded at coastal sites in large numbers, often calling excitedly and showing every sign of being 'on the move'. Subsequent ringing returns show that even these birds are of very local origin and thus, although they still retain the urge to migrate, the British birds do not actually move. On the Continent, in France and Holland, many fewer Dunnocks are ringed but there have been regular medium distance recoveries of several hundred kilometres. Further east and north the German and Scandinavian populations are regular migrants to southern Europe often covering more than a thousand kilo-

metres. Studies of these different populations have revealed that the British birds have the most rounded and the German and Scandinavian ones the most pointed wings – the Dutch and French birds' wing shape is intermediate.

These examples have shown the effect of having to make the long migration flight on the physical characteristics of the birds' wings. Earlier (in chapter 6) the different pressures on birds brought about by their annual changes in habitat and the obvious acquisition of different sorts of feeding habits – as shown by their bill adaptations – were discussed. This is borne out by the remarkable reversal of two of the 'avian laws' – Bergmann's and Gloger's Rules – pointed out by Finn Salomonsen for the Ringed Plover. Bergmann's Rule states that, within a species, size tends to be larger in the colder parts of the range and smaller in the warmer areas. Gloger's Rule concerns the general colour of a species and states that it will be more heavily pigmented in the warmer and damper parts of a species range and paler in the colder and drier areas. Salomonsen measured and looked at the colours of Ringed Plovers from the breeding populations of four areas across Europe: England, Denmark, Southern and Arctic Sweden and Russia. The palest and largest birds bred in Britain whilst the darkest and smallest bred in the Arctic areas. Apparently the two rules had been broken but Salomensen then drew a

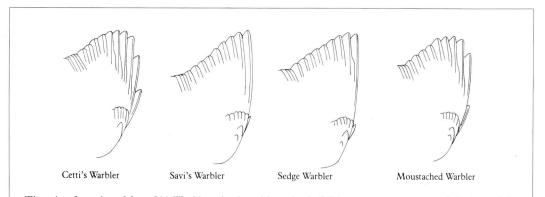

Cetti's Warbler   Savi's Warbler   Sedge Warbler   Moustached Warbler

The wing formulae of four Old World wetland warblers clearly fall into two groups – rounded wings of the comparatively sedentary (or totally resident) Cetti's and Moustached and pointed wings of the highly migratory Savi's and Sedge.

After Svensson, L. (1975)

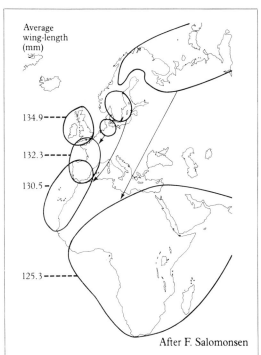

Average
wing-length
(mm)

134.9 ----

132.3 ----

130.5 --

125.3 -----

After F. Salomonsen

In winter the smallest Ringed Plovers are found
furthest south – a classic demonstration of
Bergmann's rule – but during summer they are
the 'wrong' way round (smallest birds furthest
north). This shows that the major pressures on
this species act during the winter rather than in
the summer when, in most areas, there is a
super-abundance of food for the breeding birds.
Breeding areas and corresponding wintering
grounds are joined by arrows.

map with the wintering areas of the four
populations shown (see the map above) and
all was explained. The two rules were not
being violated; it was simply that they were
operating on the Ringed Plovers in the wintering quarters and not on their breeding
grounds.

Migrants and their ranges

Some of the mechanisms by which birds have
developed into migrants have been explained.
The extension of these ideas to individual
species can be used to explain some of the
more bizarre movements performed by
present-day populations. For example, the
parental stock of the Wheatear living today
originated from migrant birds moving between Eurasia and Africa. As the breeding

birds moved northwards and westwards with
the large-scale climatic improvements following the latest glaciation, the species spread
across the north of Asia and eventually
pioneered Alaska. This is a remarkable feat
for an individual Wheatear but perfectly
reasonable for a species over thousands of
years. In fact, if the weather has been warming for the last 10,000 years the rate of spread
of breeding Wheatear need only have been
800 metres a year for it to have colonized
Alaska from a starting point in Asia Minor.
This is modest compared with the recent
spread of such birds as the Collared Dove
from Turkey to Northern Scotland in 80
years (2500 kilometres) or the rapid invasion
of the Cattle Egret into the Americas over the
last 50 years.

These considerations can explain the extensive ranges of many birds which persist in
migrating to their ancestral wintering grounds
using a route which has evolved in the same
way as their range extension. There are also
good theoretical grounds for this ultra-
conservative behaviour: range extension,
over a few kilometres, will not violate the
genetically controlled standards bred in the
species. A change of migration route, imply-
ing a shift in orientation and possibly a
different journey length, would require the
genetically controlled instructions to be
changed. As we have seen from Peter Berthold's work on the different populations of
Blackcaps, this information seems to be controlled from several loci. It is very difficult to
see how a new pattern of migration could
spontaneously arise within a population that
did not gradually *evolve* from that which
already existed. The evolution of sub-specific
differences in structure and migration patterns has been demonstrated for a wide
variety of birds. Possibly the classic example
is the Fox Sparrows of the western seaboard
of North America. The map opposite
shows how the different races have developed
in the six coastal areas and how they have
evolved their movements into a classic 'leap-
frog' format. The position is simplified in the
map, for further sedentary races of the same
species live in many parts of the winter range
of the migratory subspecies.

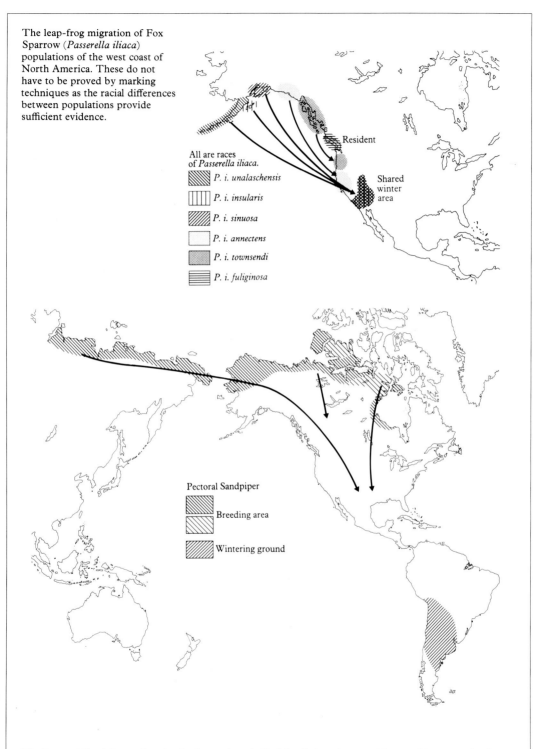

The leap-frog migration of Fox Sparrow (*Passerella iliaca*) populations of the west coast of North America. These do not have to be proved by marking techniques as the racial differences between populations provide sufficient evidence.

Resident

All are races of *Passerella iliaca*.

P. i. unalaschensis

P. i. insularis

P. i. sinuosa

P. i. annectens

P. i. townsendi

P. i. fuliginosa

Shared winter area

Pectoral Sandpiper

Breeding area

Wintering ground

The Pectoral Sandpiper, wintering in the southern half of South America, provides a striking example of range expansion in the summer – well into northern Siberia – whilst retaining its American wintering ground.

The role of migration in range extension

It is perfectly clear that even the most sedentary birds move during their lives. If they did not then there would be in-breeding at each point within the species' range and the genetic basis of the species would break down and, inevitably, genetically induced problems would cause widespread extinction. There is therefore, even for sedentary species, every possibility of gradual range extension into new areas. These may become available through changes in climate, gradual adaptation of the bird to exploit different resources or even man-made changes to the habitat. It is, however, difficult to envisage the colonization by a very sedentary species of a newly available and suitable area 50 or 100 kilometres north of its existing range, particularly if a mountain range lies between two areas of valley habitat. In fact most species which are normally sedentary do move, as young birds, over distances of this order even if their parents, once they have settled to breed, never leave their breeding territory. This phenomenon – post-juvenile dispersal – must not be confused with migration as it is generally an un-oriented movement which becomes most noticeable on the edges of a species' range.

Migrant species have the same sort of post-juvenile dispersal phase but, probably because they are adapted for longer distance flights, it may take them hundreds of kilometres from the breeding areas. Indeed in some species the post-juvenile dispersal can give rise to very large-scale movements, since the dispersing youngsters may not be orienting their flights and can easily be drifted by the wind. If they are carried out to sea, large numbers may be found on islands or in coastal areas. In many species this post-juvenile dispersal phase is difficult to distinguish from the 'familiarization' flights around the area where they will hope to return to breed. In fact there is no reason why the same flights should not fulfil both of these functions. Less easily explained are the long-distance but regular movements of young birds of a few species in the opposite direction from their normal routes. The Pallas's Leaf Warbler from the east Asian breeding area is now regularly recorded in the late autumn in Western Europe. It has been argued that

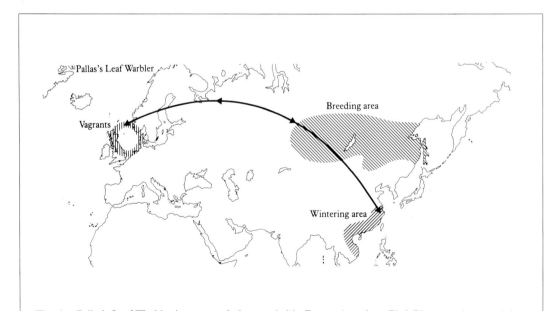

The tiny Pallas's Leaf Warbler is now regularly recorded in Europe (mostly at Bird Observatories round the North Sea). These birds (always youngsters) may well have a defective compass for they would appear in Europe if they migrated on the reciprocal of the 'correct' bearing of their winter quarters.

these birds are making a simple mistake of 180 degrees in their migratory flight, since the Great Circle route projected from Britain through their breeding area for the same distance beyond reached their wintering grounds. Whether any survive to return to Asia is not known and, although the birds that are caught are ringed and released, many centuries of marking will be needed before this question can be answered by conventional banding techniques. If wireless transmitters for use on birds are further miniaturized, becoming a tenth their current size *and* several thousand times more powerful, the question may be answered in one or two seasons.

In any case, it is very easy to see how new areas may be pioneered by young dispersing migrants over quite long distances. There are, however, many other ways in which new breeding grounds may be colonized by fully-grown migrants. Some may remain in their wintering areas, if the conditions suit them or if they are inhibited from making the spring migration. Reg Moreau found that some 20 per cent of all European migrants to Africa also had breeding populations south of the Sahara. In addition, a further twelve species had been known to nest in African wintering areas but have not regularly established breeding populations. This phenomenon is also well known in the Americas with, for example, the scattered breeding populations of American Bitterns within the wintering area. In many cases colonists of the wintering areas have gradually developed their own characteristics and become at least racially distinct. This has happened to the Broad-winged Hawk on the Caribbean islands with five sub-species and a fully distinct species now in existence. Charles Leck, who recently reviewed the evidence for this sort of range extension, found further examples in American cranes, geese, ducks, rails, waders and owls. Similar extensions of breeding range might also result from birds stopping off on spring, or less likely, autumn migration.

There is always the possibility of birds which have become lost on migration pioneering new sites. Two circumstances are particularly auspicious. The first is during the spring if there is fine weather and numbers of birds of the same species over-shoot their normal summering area. This often results in isolated pairs of birds breeding to the north of their species' normal range. If they are successful then a population may become established. Similarly very bad weather in the spring may retard the migration and lead some birds to settle and breed further south than normal. The other favourable circumstance is on oceanic islands where, if two birds of opposite sexes happen to meet, and there is suitable habitat, breeding may result in colonization and swift evolution to form new races or even species. The effects, as demonstrated by Darwin for the finches on the Galapagos, can be staggering with successive colonizations by the parent stock resulting in very different species evolving. This can only happen when the new invasion is long enough after the earlier one for the original colonists to have changed enough to ensure reproductive isolation from the newcomers.

Man-induced changes

It is very easy to forget that the world we live in is very much the product of man's modification of the natural world. In the temperate zones even the apparent wilderness areas have probably altered because of human influence over the last 500 years. In existing areas of woodland there has often been much felling of timber with consequent gross changes to the ecosystem. For example, it has been calculated that true primeval woodland contains 50 per cent more dead wood. The fallen trunks and branches consequently contain increased populations of wood-eating invertebrates and of holes and crevices for nesting and roosting birds. Many areas of open moorland are the direct result of human exploitation – often the clearing of natural woodland then followed by some farming activities which deplete the nutrients in the soil.

In areas of high human population the 'natural' habitat for land-birds has largely disappeared to be replaced by towns and villages, arable land and grassland, orchards and planted woodland. The birds which are able to exploit these altered habitats include not only many highly adaptable species but also specialists from particular niches which happen to approximate to the conditions which man has made. The Swallow (known as Barn Swallow in North America) provides an excellent example of this latter strategy. There can be few ornithologists who have seen a 'natural' Swallow's nest – presumably they used to build nests in rock caves or even hollow trees – but everyone will have seen nests in buildings. Such nests were certainly being built in Europe during Roman times. It is very likely that the buildings, as potential nest sites, enabled Swallows to colonize large areas from which they used to be excluded through the lack of nest sites. In Europe the Common and Pallid Swifts are further aerial feeders whose nests are now almost exclusively built in buildings; in America, the same is true for the Chimney Swift, a species which now inhabits vast tracts of land.

The position of species which were probably native breeders in the pre-human era is most interesting. The habitat now available provides little opportunity for specializing species which were dependent on particular trees or large areas of woodland. However, the more flexible woodland species are now able to use many areas which, to the human eye, do not really look like woodland. For example, suburban gardens are a good approximation to woodland and are colonized, in Britain, by wood-loving species like the Great and Blue Tits, the Song Thrush, the Blackbird and the Chaffinch. In areas with small fields, whose boundaries are hedged with trees and scrub, these woodland birds are now able to live with scrubland birds, and even open country birds in the fields. Such an area might have Willow Warblers, Whitethroats and Linnets (scrubland species), the woodland ones already mentioned and Skylarks, Lapwings and Partridges as examples of field birds. These changes will have had enormous effects on the breeding populations of summer migrants and also of winter visitors. Until fairly recently the tidal habitats used by waders

have not been altered much by man but, in recent years, large-scale civil engineering projects have altered that. For many centuries the extensive drainage of freshwater marshland areas has greatly changed the patterns of distribution of ducks, geese, waders and other waterbirds.

These effects are not only apparent in the main temperate areas but also in sub-tropical and tropical areas. In some instances the changes have been on a very large scale and are the result of alterations to the climate which have come about apparently from man's activities. For example, the extension of the Sahara Desert over the last 2000 years has been related to farmers clearing areas of original vegetation to grow annual crops. Slash and burn methods have been practised for thousands of years in parts of the tropics. There is now good archaeological evidence to show that several different civilizations that used this method flourished in South America before the conquest. The human populations involved were numbered in many millions and their activities will probably have altered the South American ecosystem much earlier than the Red Indian or European settlers had any material effect on the North.

It seems significant that studies in both Africa and South America have shown that the majority of the small long-distance migrants from further north tend to concentrate in modified areas where man's influence is obvious. These may include the areas being actively farmed but seem particularly to include areas of regeneration following clearance. What seems to have happened is that the ancient habitats have evolved complex communities of plants and animals into which the local birds fit securely as specialists themselves. When this balance is upset by human activities the conditions which result are best exploited by the more generalist migrants from further north.

Migrating seabirds have probably been least affected by human activities. This is now changing. The pressure of large-scale fishing on the higher trophic levels within many of the world's oceans has caused huge changes in the numbers of fish present. The migrants are most often affected on the breeding grounds and have, in many cases, profited greatly until now. This is generally because the human exploitation of the bigger, more predatory fish has enabled the lesser species, like Sandeels in the Atlantic, to increase tremendously. These form the preferred food of many seabirds whose populations have consequently increased. As the larger species have been fished out man has started to exploit these lesser species for fish-meal and crude protein in direct competition with the birds. Off western South America a similar situation arose about 10 years ago with the local fishermen beginning to take very large amounts of fish for fertilizer rather than exploiting the guano from the seabirds. The seabird populations there have suffered to a very considerable degree.

The future

Obviously the numbers of seabirds may be on the wane within the next few decades as fisheries compete with them for food. In the same way, if extensive tracts of temperate region estuaries and mud-flats are to be reclaimed for farming purposes or embanked for water storage, profound changes either in migration patterns or in numbers are bound to affect many species of waders. Luckily this is now becoming understood not only by birdwatchers and conservationists but also by the public at large and even some politicians. Provided that organizations like the Royal Society for the Protection of Birds and the Royal Society for Nature Conservation in Britain and the Audubon Society in America are able to carry on their programmes to cultivate this public awareness and translate it into legislation northern and temperate areas should retain extensive areas of these habitats. Further south, in the developing world, the pressures may be much greater but many countries are already showing an awareness of the importance of natural habitats as part of their national heritage, as well as a significant tourist attraction.

On the land the pressure to drain freshwater marshes and to assist with the flow of water through river systems – or to store water in reservoirs – will persist in all areas. These may alter, as they have over the last centuries, the movement patterns of many species. The introduction of different agricultural methods, ever geared to more efficient production, may provide new opportunities for migrant birds to feed. These are likely to be few, for the object of agriculture is to provide the correct conditions for monoculture. This unnatural situation seldom operates to the bird's benefit: the exploitation of the wheat and rice fields in the Niger inundation zone by Gargeney and Ruff being a notable exception. Agricultural practice is becoming more industrialized throughout the world and this means that there are few places where pesticides are not in regular use. Migrant birds are at particular risk since they may become contaminated in areas which they may only visit for a few weeks of the year. Once more it is likely to be the emerging nations of the third world, where famine is a real threat, that may over-enthusiastically use pesticides injurious to birds. However, the birds themselves, if contaminated and dying, act as a very salutary warning to the human population. They are unlikely to suffer for very long before connection is made by the local inhabitants!

Over the longer term, migration patterns will continue to alter as climate alters. In the future this is likely to be as a result of human manipulations just as much as natural changes. The proposal by the Soviet Union to alter the flow of some of their major rivers may have an effect. Another factor could be the accumulation within the atmosphere of propellant from aerosols and of carbon dioxide from the burning of fossil fuels. Within smaller areas the 'thermal pollution' of lakes, rivers and the sea by waste heat from power generation may provide special habitats for birds which would normally move further south. There are also predictions, based on the theory of plate tectonics (Continental Drift), that in 50 million years there will once more be a sea gap between North and South America which may inhibit migrants from making the crossing. It also seems likely that Australia will have moved much further north and may have a part to play in the migrations of many more Palearctic landbirds than it does at present.

183

Investigating migration

All the theories, facts and figures so far presented, and everything that will ever be found out about migration, is based on research.

The research methods used may simply be unaided observation of the birds of an area by a local birdwatcher, be it Aristotle in Greece thousands of years ago or a present-day member of the Royal Society for the Protection of Birds or the Audubon Society. They may, however, involve specially designed and very sophisticated technology. It is not surprising that the major advances in our knowledge of the mechanisms of navigation and the physiological control of migration now mainly stem from professional research teams using high technology. However, there are many other interesting means of investigating migration which can still be used without investing in more than a decent pair of binoculars.

Indeed, useful results can even be obtained by research workers who need never see a live bird but rather study preserved skins in museums or even examine the corpses of freshly shot birds! This chapter is devoted to reviewing the research techniques that have been used and are in use at the moment.

Museum studies

The great museums of the world contain literally millions of preserved skins of birds. In most parts of the world new specimens are being added each year. In Britain new specimens come mainly from bird casualties found on roads, under windows or at lighthouses. In the Victorian age an enormous amount of bird-shooting was done for 'preservation' – that is the shot birds to be made into study skins. Many birdwatchers today feel that shooting bird specimens for skins is cruel and unnecessary. Certainly current skills in field identification have proved the lie of the old adage that with a rare bird 'Whats hit is history, whats missed is mystery'. An accurately taken field description by a competent ornithologist will serve to identify almost every species likely to be found in Europe or North America. If further proof is needed photographic techniques allow a permanent record to be taken without harming the bird, particularly if it has been caught for ringing purposes.

However, this is the situation now only through the work over the last 300 years or so of generations of ornithologists who did kill birds to make specimens. From their specimens they were able to work out with precision what the different species looked like and to publish detailed descriptions. All our birding nowadays is based on this knowledge. In many of the more remote areas of the tropics in South America and Africa, collecting is still providing this basic knowledge about new species and races of birds. In some notable cases, the greatest care is being taken to ensure that there is no unnecessary killing of birds. One African museum regularly uses mist-nets to catch birds. From the birds caught those of no interest are immediately released, those of some interest are photographed under standard conditions (having been administered a light sedative) and are then released; only the birds not fully represented in the national collection of skins are killed and preserved.

Museum studies form the basis of our ability to 'name' the different species of birds but they can also provide direct information on the migration of many species. When the 'bulk' collectors of birds worked an area of South America or Africa they will have shot and trapped all the birds that were 'shootable' or 'trappable' in the area. Their aim was generally to preserve as representative a sample of the localities avifauna as possible. These 'samples' inevitably included migrants from further north, if they were taken in the winter. This is how the wintering areas of many species were first established. It is easy to think of this as being only an historical technique but it can still be used to advantage nowadays. A recent example is David Seel's work on Cuckoos in Europe and Africa. The Cuckoo is a conspicuous bird which is an easy target for a collector but is very difficult for bird-ringers to catch. David has been able to map its migration simply by reading the labels on the specimens in large numbers of museums and recording the region where the Cuckoo was 'caught' and the month of the year. The map (right) shows Cuckoo specimens from west of 60°E for June and December. Because the birds involved were museum specimens it was always possible to distinguish them from the very similar African Cuckoo.

Similar to these museum investigations are studies resulting from the mass mortalities of migrant birds. In New Zealand a great deal has been learnt about the movements of seabirds by the regular patrolling of beaches to find and preserve corpses (or at least the wings) of dead birds. Stranded corpses account for a number of vagrant records in Europe and America, although there are always slight doubts even about fresh ones. For instance, a fresh penguin corpse found in Kent was eventually identified as a casualty from a consignment destined for a zoo. Lighthouses, whose beams attract migrant birds, provide a certain but sad way of logging migration. In many cases such navigation lights do not prove a fatal attraction but there are lights, like the one on Bardsey Island off North Wales, where birds may follow the flashes made by the rotating beams round and towards the light, eventually crashing into it.

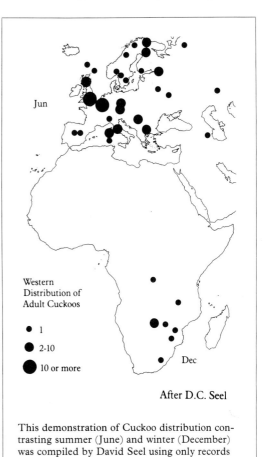

Western
Distribution of
Adult Cuckoos

● 1

● 2-10

● 10 or more

After D.C. Seel

This demonstration of Cuckoo distribution contrasting summer (June) and winter (December) was compiled by David Seel using only records of specimens in major museum collections.

Above Birds attracted to the rotating beams of Bardsey Lighthouse in North Wales. Heavy attractions, involving thousands of birds, generally coincide with misty weather at new moon periods during the major migration seasons. *Left* The night's toll at Bardsey. Many of the birds may have killed themselves on the glass of the lighthouse lantern.

Many of the birds in the beam will simply flutter round but there are sometimes hundreds killed in a single night in the early spring or late autumn when many of the migrants are larger species like Starlings and Redwings. There are a number of species, like Snipes, Water Rails, Sedge and Grasshopper Warblers, which are notoriously prone to such accidents. The authorities in England and Wales, including Trinity House and the Royal Society for the Protection of Birds, have co-operated to reduce such hazards. On Bardsey, the lighthouse tower is illuminated by floodlights if birds are being attracted. An adjacent patch of gorse is also lit as a sanctuary for grounded birds.

Inland in North America there have been

numerous reports of very considerable kills of birds against high man-made objects. These are generally TV towers and the numbers of birds killed during a migration season at a single tower have exceeded 1000 at many places. Systematic collection and recording of these corpses has provided very useful information on the timing of migration and the relative abundance of similar species. It is thought that the deaths are a representative sample of the birds flying, on any particular night, at the height 'swept' by the tower. Species like waders, ducks and geese, which normally fly several thousand feet up, are not included in the sample; neither are diurnal migrants which are generally able to see the obstruction and avoid it. Tall buildings, especially those with lighted windows, have also produced migrant casualties at night. At some developments in North America so many casualties have resulted that cleaners are employed to sweep up the bodies at the base of the buildings. Such casualties, like those at lighthouses and, to a lesser extent, those at TV towers, are most likely during overcast and foggy weather. The most extensive use of such data was made by Ian Nisbet. When investigating the frequency of Blackpoll Warblers on autumn migration, he recorded in his 1970 paper deaths of more than 140,000 warblers from 135 sites in the eastern part of North America: 5.2 per cent of these were Blackpolls.

Field recording

The rather morbid recording of dead birds chronicled so far is not the usual pursuit of birdwatchers. They are far more likely to be seeking rare birds using the modern and well-developed 'grapevine system' which allows them to pass the latest information very quickly indeed. Such 'twitching' (British term) or 'listing' (American) can be great fun but does not have much to do with the systematic recording of bird migration. The dedicated migration recorder normally chooses to count the number of birds over a particular area each day (or each weekend) so that fluctuations in populations can be traced. Some of the earliest of these records come from the diaries of eighteenth and nineteenth century clerics where, although the birds were often not counted, the first dates on which migrants were seen and, much more difficult to record properly, the last dates before their departure were entered.

Following the pioneering work of skin collectors like Eagle Clarke and the Duchess of Bedford on Fair Isle in Scotland, birders began to realize that headlands and small islands were really good places to watch birds. Given the right weather conditions, not only would rare and interesting species of birds be seen but also there might be very large numbers of grounded common migrants. Ronald Lockley, on Skokholm off Pembrokeshire (now Dyfed) Wales, started the first of the modern Bird Observatories before the Second World War. An integral part of such observatories is their ringing activities but for the moment we concentrate on the ordinary bird observations. In most cases there have been daily census records over each migration, and often for most of the rest of the year, since the observatory was opened. These assessments are arrived at through the recording carried out not only by the warden but also, in most cases, by the visitors staying at the observatory. They are usually made, often after much discussion, during an evening 'call over' when all who were watching take part. It may seem an easy task to do such a count over a small area of fields and bushes but the birds do their best to make it very difficult indeed. Some species skulk and are almost invisible whilst many tend to move around a great deal when they are grounded on a small island. It is not unusual for a real dawn count of 60 or 70 Wheatears to have diminished to 5 or 6 by the late afternoon; the rest had filtered off the island or found more congenial areas elsewhere on it. Nonetheless, these census results provide a very interesting series of quantitative records spanning decades. Their results are of most interest in quantifying the migration periods of the different species and in showing the effects of weather. This is of great interest but also has been shown, mostly by radar observations, to be the biggest drawback of the bird observatory results. The birds recorded were mainly migrants which had made mistakes, often induced by bad weather. If the migration goes well during a season the observatories tend to record very few birds, as they are all overflying the off-shore islands and headlands. If there is continual bad weather then vast numbers of birds may be recorded.

The very fact that bird observatories and other islands and headlands tend to be the first landfalls for lost migrants means that they have much more than their fair share of rare birds. True vagrants thousands of miles from home are much more often reported from such places than in 'the wood down the road'. Regular observations have also shown that many species which used to be thought

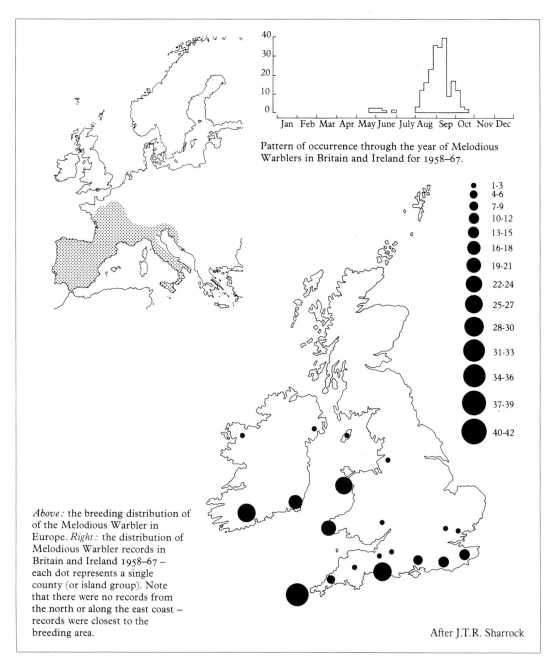

Pattern of occurrence through the year of Melodious Warblers in Britain and Ireland for 1958–67.

•	1-3
•	4-6
•	7-9
•	10-12
•	13-15
•	16-18
•	19-21
•	22-24
•	25-27
•	28-30
•	31-33
•	34-36
•	37-39
•	40-42

Above: the breeding distribution of of the Melodious Warbler in Europe. *Right:* the distribution of Melodious Warbler records in Britain and Ireland 1958–67 – each dot represents a single county (or island group). Note that there were no records from the north or along the east coast – records were closest to the breeding area.

After J.T.R. Sharrock

of as irregular visitors are actually regular passage migrants. Good examples are the related Icterine and Melodious Warblers and their occurrence in Britain and Ireland. Up to 1937, the *Handbook of British Birds* (published in 1938), lists five Melodious and 48 Icterine Warblers. Tim Sharrock, in his review of scarce migrants covering the years 1958–67 records sightings of 217 Melodious and 311 Icterine. Their distributions show a fairly similar seasonal pattern. The locations where they have been recorded are very different, with the Melodious clearly a bird of the south and west and the Icterine of the east and south.

The other birdwatching activities at bird

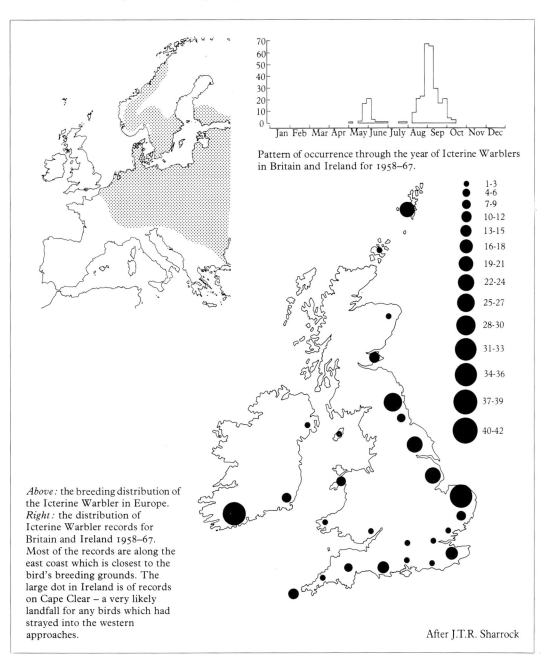

Pattern of occurrence through the year of Icterine Warblers in Britain and Ireland for 1958–67.

1-3
4-6
7-9
10-12
13-15
16-18
19-21
22-24
25-27
28-30
31-33
34-36
37-39
40-42

Above: the breeding distribution of the Icterine Warbler in Europe. *Right:* the distribution of Icterine Warbler records for Britain and Ireland 1958–67. Most of the records are along the east coast which is closest to the bird's breeding grounds. The large dot in Ireland is of records on Cape Clear – a very likely landfall for any birds which had strayed into the western approaches.

After J.T.R. Sharrock

189

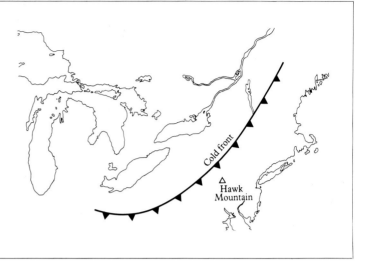

Hawk Mountain in Pennsylvania is a classic site for watching bird of prey migration. It is sited on a ridge between the Great Lakes and the Atlantic coast. The soaring raptors, such as the buzzards and broad-winged hawks, will avoid these large areas of water as they provide no thermals and the birds are forced to employ flapping flight for which they are ill-suited.

observatories are concerned with birds which are actively migrating. These may be diurnal migrants, generally flying along the coast within an hour or two of dawn or, less commonly, long-distance nocturnal migrants making a landfall. Otherwise they are likely to be seabirds flying to and fro at sea. Good movements of diurnal migrants may usually be seen when the wind is *against* them for, through the friction of the air on the land, the wind blows less strongly at lower heights. If the wind is in their favour the migrants, even diurnal ones, may be flying well out of sight to the naked eye at heights of 1000 or more metres. Sites where such movements are most regularly seen are generally in special positions where the local topography tends to channel the birds (see maps on page 191).

Watching passing seabirds can easily become an obsession. I was once asked whether I was feeling well by a man walking his dog along a Sussex beach during a January blizzard; he happened to pass me twice and on both occasions I was fully immersed in my waterproof sleeping-bag getting the circulation going between ten-minute stints of watching the sea (no birds!). It can make for very exciting birding if you have a good day but it is very difficult to work out what one's results mean since the seabirds are such mobile creatures.

The unfortunate effects of weather on birds at coastal sites has lead to many investigations of migration inland. These have included many sterling individual efforts, like the daily observations in Regent's Park, London

Average number of major seabird species seen flying west during one hour of sea-watching in each month at Cape Clear, Ireland. These figures are based on 2952 hours of systematic sea-watching during 1959–69.

	Jan	Feb	Mar	Apr	May	Jun	Jul	Aug	Sep	Oct	Nov	Dec
Fulmar	71	17	19	39	20	16	45	77	16	★	11	34
Manx Shearwater	1	2	60	213	204	249	758	369	37	★	★	★
Great Shearwater			★			★	3	1	18	★		
Storm Petrel			★	★	★	★	5	30	4	★		
Gannet	76	49	114	95	111	108	164	298	309	245	44	30
Kittiwake	82	25	44	46	56	35	19	22	19	32	57	35
auks	141	186	238	373	179	76	14	1	17	67	131	57
Total	371	279	475	766	570	484	1008	798	420	344	243	156

★ = Average of less than 1 an hour

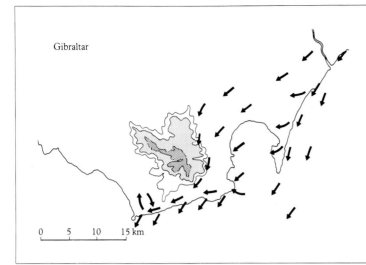

The short sea-crossing between Gibraltar, at the tip of the Iberian penninsular and Cape Spartel in Morocco, at the western end of the Mediterranean, is one of the best places in Europe for watching the migration of larger raptors and other soaring migrants, such as storks. If the weather conditions are unfavourable for a crossing the migrants build up in large numbers over the Spanish mainland and then cross in their thousands when the weather improves.

After Bernis A.J. *et al* (1973)

Map of the British Isles showing the position of Cape Clear Bird Observatory. This observatory is superbly located for watching the passage of seabirds making their way along the eastern Atlantic seaboard.

carried out for several years by Ian Wallace and his friends. This unlikely locality provided many records of arriving and passage migrants. Its great improvement on the coastal observatories was that the arrivals could actually be correlated with mass movements of birds, even when the weather was so good that none had become grounded at the coastal sites. The most successful investigations have actually relied on large numbers of observers sending their records into central sites. Notable research of this sort has been

the maps of isochronal lines which link the dates of first arrival of summer migrants over Europe (Southern) and North America (Lincoln). More ambitious schemes with the observers sending in census counts for small areas throughout the country have been tried in the past. The Inland Observation Posts Scheme (later the Daily Bird Count) had more than 100 observers but, with each recording an average of 30 species a day, the data soon overwhelmed the part-time organisers. Even this modest scheme could produce over a million figures per year and, when it was being tried, computers were still a thing of the future.

The efficient collation of records from many sources can provide for quite good descriptions of migration patterns. Once more taken from Tim Sharrock's study, the maps and histograms of Hoopoe records (see page 192) in Britain and Ireland result from a thorough search of all the county and regional bird reports published over the years 1958–67. Many of the observers of these birds would not have realized that they were possibly contributing to a serious study of the bird's movements but rather simply felt that they should record the relatively rare bird they had seen. Similar careful searches through the local reports have produced useful analyses of the movements, inland, of several coastal wader species. Indeed there have been, in several countries, specially

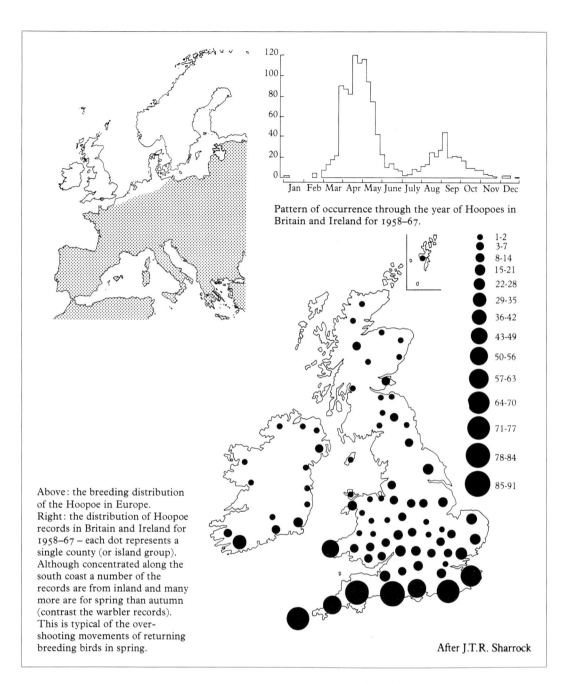

Pattern of occurrence through the year of Hoopoes in Britain and Ireland for 1958–67.

1-2
3-7
8-14
15-21
22-28
29-35
36-42
43-49
50-56
57-63
64-70
71-77
78-84
85-91

Above: the breeding distribution of the Hoopoe in Europe.
Right: the distribution of Hoopoe records in Britain and Ireland for 1958–67 – each dot represents a single county (or island group). Although concentrated along the south coast a number of the records are from inland and many more are for spring than autumn (contrast the warbler records). This is typical of the over-shooting movements of returning breeding birds in spring.

After J.T.R. Sharrock

organized investigations which have involved daily co-ordinated counts of migrant waders on such favoured inland sites as sewage farms, settling tanks and gravel pits. The inspiration for these can be traced back to David Lack's pioneering observations, made whilst still an under-graduate, at a Cambridge sewage farm.

Marking the birds

Banding (American term) or ringing birds with numbered metal rings bearing a return address has now been in regular use for more than 80 years. As a technique its great advantages are that the exact whereabouts of the bird is known twice in its life and the finder need not be able to identify the bird –

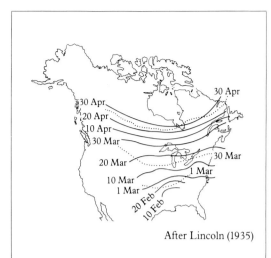

After Lincoln (1935)

The isochronal lines (lines joining places with first records on the same dates) of Canada Geese in spring (solid lines) which move north at about the same time as the 35°F isotherm (dotted line).

of all birds which are ringed nowadays are caught as free flying individuals in various traps and nets (see photos on page 194). In many countries the methods shown are illegal unless used for ringing purposes by a fully trained and qualified ringer.

The finder of a ringed bird should always report it, even if it is known to have been recently and locally ringed, as this is the way that information can gradually be pieced together. The ringed bird, if it is alive, should be looked after humanely and released with its ring still on, after the number has been carefully *read and written down*. With dead or hurt birds it is advisable to remove the ring and tape it to the letter reporting the bird; in such cases please write the ring number on the letter too. All ringing centres have stories to tell of thrilling letters reaching them from distant lands in envelopes with little holes from which the ring has fallen in the post. The information in the letter should be simply:

Date the bird was found.

Where it was found so that the office can pinpoint the spot on a map.

What had happened to the bird (shot, road accident, cat victim or plain 'Found dead') and was it fresh or not.

Name the species or type of bird.

Who you are so that the scheme can tell you the ringing details of the bird you have found. All schemes do this for all finders of their birds.

In most cases the address on the ring will be understandable by the post office in any country but it may save time and expense to send the information, in the first place, to your national ringing office. For one thing there will be no difficulties in translating from one language to another.

In recent years many ringers have co-operated to ensure that there is more chance of finding out about their ringed birds other than the remote possibility of a bird being found by a member of the public. They have been able to organize co-operative enquiries designed to catch each others birds at different places. The two most famous schemes have been Operation Recovery on the eastern seaboard of America and Operation Baltic on

the ring number is sufficient. Unfortunately there are grave disadvantages since only a minute percentage of small birds are likely to be reported. In well populated areas this is typically about 1 per cent but in areas with sparse human populations literate enough to read the ring it approaches 0 per cent. Nevertheless, the grand total of ringed birds, worldwide, must now be approaching 100 million, with an increase of 4–5 million per year. The national ringing schemes co-operate fully so that information passes between countries without too much difficulty; third parties are sometimes used to make the connection between countries whose politics allow no direct contact.

The catching and marking of the birds may be done by professional or amateur ringers but all work is to a high standard. It is totally pointless to mark birds which have been damaged in the catching process or to use bands which might harm the birds since the basic point of it is to investigate the behaviour of fully healthy and unrestrained wild birds. Many birds are ringed in the nest or in colonies before they are able to fly. Subsequent records of these recoveries are particularly valuable since their exact age and origin is known. However, the vast majority

the southern coast of the Baltic. In Operation Recovery ringers manned net sites along the coast and were quite successful in exchanging birds. In Operation Baltic the best results came from winter visitors moving westwards in the autumn. These involve many ringers keeping up their efforts over a period of weeks and for several years. Other teams have gone to the areas where they know, from previous recoveries, that their migrants are passing. This technique has been very successful for waders in America, Europe and North Africa. Other expeditions have included the famous northern trips by Peter Scott to the breeding grounds of geese, such as the Pink-footed in Iceland. Marking them there was

Above left This Linnet is being released from a very fine net – a mist-net – by a trained ringer. These nets are ideal for ringing small birds since they are portable and quickly erected. Their use, over most of Europe and all North America, is illegal without a government licence.

Left A Fieldfare being removed from the catching box of a Heligoland trap at Sandwich Bay in England.

Below Traps like this, placed at the water's edge, are baited with grain in order to catch ducks for ringing purposes.

an excellent strategy since they were bound to provide winter recoveries in the areas where they were regularly shot.

For many years ringers have been dissatisfied at the rather low recovery rates of the long-distance migrants and so they have sought to get as much information as possible from the birds they handle. The age and sex of the birds are carefully recorded as well as their species. They are weighed and measured in the hope that they are later retrapped so that the rate of weight-gain for migration can be measured. Special ringing expeditions have been mounted to areas of interest where the main object was to discover the weight regime of the migrants and the length of their passage period. The study of inland weight gain of the Blackpoll Warbler in New England was the classic example which spawned many imitators, such as the expeditions to France, Spain and Portugal by British ringers in the 1960s and 1970s.

From such investigations other projects followed. Ringers with the chance of marking the summer visitors to the northern temperate areas whilst they were on their wintering grounds realized that they stood a much better chance of long-distance recoveries. In this field the regular banding of several thousand small migrants late every autumn at the Ngulia game lodge in Kenya deserves special mention. The small group of ringers, led by Graeme Backhurst and David Pearson, have ringed there regularly each year since 1973. The lighted windows of the lodge, facing north, act as an attraction point just like a lighthouse, since it is on the edge of north-east facing high ground. Migrants

Below A male Blackcap having its uniquely numbered ring fitted to its leg.

Bottom A double-ended 'Wall' trap on Fair Isle, Scotland. Birds are gently moved along the wall and are caught in the large funnel of netting and are extracted from the catching box at the end. Such traps are called Heligolands, after the German island where the technique was introduced.

actually fly into the open rooms of the building or are caught in the compound outside – not too far outside because of the big game!

Another method of getting better results from ringing is to mark the birds with something that can be seen and identified in the field without having to catch and handle the bird again. This can be achieved in many different ways: with colour rings on the legs; large rings with numbers, letters or patterns that may be read through binoculars or a telescope; back harnesses with tags; numbered tags attached to the wings; and dyed plumage. All have been tried with varying success but they need to have international co-operation and forward planning. It would be disastrous if a Danish expedition to Greenland dyed Ringed Plovers orange (colour produced by picric acid, a dye which is very long-lasting as it chemically combines with the feather material) and then returned to find that an English group in Iceland had done the same! An advanced field technique used on Bewick's Swans relies on the individual recognition of the bird's variably coloured and marked beaks – really a case of 'self marking'.

The most recent marking technique involves the attachment of tiny radio transmitters to the bird so that they send out a signal which can be picked up by the research worker. These are generally used for studies of birds which are likely to stay in one area because the range of the miniature radios is not long. However, migrants have been traced successfully over several hundreds of kilometres of flight by mounting the receiver in a fast car or onboard an aircraft. As the radios get smaller and receivers more powerful, it will be perfectly possible to trace the movements of an individual bird through satellite links. This has been proposed for some time for large seabirds like albatrosses but is not yet in operation.

Birdwatching at night

The great majority of migration takes place at night and is therefore invisible. Or is it? On a clear night there is, for part of every month, an almost full moon high in the sky for some of the night. By focussing a pair of

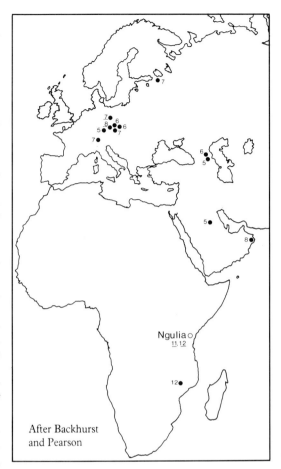

After Backhurst and Pearson

Map showing recoveries of Marsh Warblers controlled or ringed at Ngulia. Solid circles and numerals represent recovery site and month of a Ngulia-ringed bird. Solid circle and underlined numeral represents ringing site and month of a bird controlled at Ngulia.

binoculars or, better still, a lower power telescope on the disc of the moon it is possible to see, in silhouette, the birds, if any, which fly between you and the moon. This can only be done systematically if the observer is really comfortable and has the scope mounted on a solid tripod. If you know which way is north and make a diagram of the moon on which to draw any movements seen it is perfectly possible to work out what direction the birds were flying relative to the ground. Obviously the high flying small birds, which may have a span of 15 centimetres and be at a height of 2 kilometres, are likely to be missed. It is a task roughly equivalent to spotting a dime or a new penny at 200 metres! It also

This young Red Kite is marked with a coded wing-tag to enable its movements to be traced. Individual birds, marked in this way, can even be recognised in flight – through powerful binoculars!

needs a team to keep up continuous observations since one's eyes become tired very soon. Even two people working turn and turn about, lasting 10 minutes, may find it tough work.

Moonwatching data, used with information about the wind, can enable the bird's actual heading (rather than its resultant ground tack) to be calculated. In exceptional circumstances the species involved can be identified. This is generally with high-flying large birds which remain visible for several seconds as they appear across the moon's disc. Low-flying birds zip across too fast and, of course, small passerine species look much too similar for specific identification when seen in silhouette, with no reference as to their size. Moonwatching records may be as few as two or three passes an hour, even during peak migration periods in unfavoured areas. The record highs are of the order of 200 passes an hour which, dependent on the height of the birds and the altitude of the moon, may represent 50,000 birds per hour passing over a mile front!

A rather similar technique, in reverse, has been the use of a ceilometer's lights for observing birds. The ceilometer is a device for measuring and recording cloud height, based on the reflection of a beam of bright and focussed light from a cloud base. The early models used in the 1940s were sometimes responsible for killing migrants but filter devices were introduced in the 1950s that stopped this. More recently fixed-beam which are safe but, unfortunately, useless for the research worker. In order to observe what is flying through the beam the observer mounts his telescope vertically a few feet to one side of the ceilometer and up the beam. The off-set is most necessary as the lower part of the beam will inevitably be crowded with insects and also have a constantly moving cloud of dust particles in it. The birds seen flying through the beam are then recorded in the same way as for moonwatching. The interested amateur can try this by mounting a powerful spotlight vertically. Care needs to be taken in choosing one with as thin a 'pencil' beam as possible and also in not sighting it so as to blind the pilots landing at the local airfield. The vertical telescope technique can, of course, also be used to record very high-flying diurnal migrants. The density of migration can be calculated using the moonwatching tables, correcting for the difference between the field of view of the scope and that subtended by the moon.

One characteristic of many nocturnal migrants is their contact calls. These are often very different from the normal calls which birdwatchers associate with the species but they are familiar sounds of the night in spring and autumn. Unfortunately they travel different distances in different weather conditions and the birds seem to be very much more vocal in overcast and foggy weather. There have been attempts to quantify the migration going on over a season by recording the numbers of calls heard but all have failed to prove that what was being measured was the 'quantity' of migration. Two groups are of particular interest. One is the geese which in some areas are only regularly recorded as noises passing in the night. The other are the waders which may often make their own distinctive noises at night on migration. In foggy conditions, even at inland sites, it is not unusual to be able to record five or six species in the course of half an hours listening. Bird-watching without using one's eyes is equally possible in town and country, provided the traffic noise is not too great. Indeed in some weather conditions lost waders will fly round and round the lights of a town.

Radar migration watches

The first radar sets were developed for use during the Second World War and were plagued by 'angels'. These were echoes which showed up on the screens where no aircraft should be. Research indicated that these 'angels' were the scatter of radio waves sent back to the receiver from water droplets, ice particles, birds or insects in the air. By an extraordinary stroke of good fortune David Lack was put to work by the army on the operational problems of their radar network along the south-coast of England. As early as 1941, David Flack realized that radar had great potential for tracing migrating birds. However, it was not until he became director of the Edward Grey Institute that this was to bear real fruit. Radar sets are made for all sorts of purposes and some proved particularly useful for 'watching' birds. On the face of it birds do not seem to be ideal material to reflect a pulse of radio energy but within the soft casing of feathers each bird has a sphere of tissue, bone and, most important, water. This actually forms quite an efficient scattering body for the radio energy. The bird is tiny in relation to the aircraft that the radar is normally designed to track, but the signal that the radar receives from its target is only dependent on the target's cross-section through an inverse fourth power law. The

example that Eric Eastwood gives in his book *Radar Ornithology* compares the 0.002 metre^2 head-on aspect of a Starling with the echoing area of a Canberra (twin engined jet bomber) 55,000 times that size. However, because of the inverse fourth power law the signal received from a Starling is only a fifteenth that from the Canberra!

Unfortunately, an ordinary surveillance radar, designed to 'see' aircraft, does not have much resolution. This means that a single bird within the 'pulse-volume' of the set will give rise to a blip on the screen, as would 500 birds (although the blip might be brighter). The pulse-volume of a set may be a square 500 metres in cross-section and reaching far above the height at which birds migrate at a range of 30 kilometres from the set. This is equivalent of a cube with kilometre long sides! However, the great achievement of radar observations were two-fold. The first, already mentioned, demonstrated that the majority of birds migrated successfully under good weather conditions and never made landfall at the coastal bird observatories. Secondly, streams of migrants could be tracked moving in all sorts of different directions on the same night, often being very obviously influenced by weather conditions.

In the early years bird observations on radar sets were made by the ornithologist sitting watching the tube and making notes. Soon it was realized that, with slow-moving echoes of birds, the ideal means of recording was time-lapse photography. Simply set up the radar, with a watch beside it, and take shot after shot of the tube every minute or two. However, the radar picture is not very bright and eventually it was decided that a cine camera with a time exposure of 15 seconds for every frame was ideal. This compares with 25 frames per second for normal shooting and so, run through the projector at about normal speed six hours of observations would only take one minute to view. At last moving pictures of night migration were available. In fact, because of the way that the image is retained on the tube, moving objects trail faint tails behind them, in the form of after-images. Even still shots can tell one a great deal about what is happening. The scale of the picture one can observe with radar can be altered. The surveillance radars can easily give an over-view of the movements of birds over an area more than 100 kilometres across. The photograph on page 201 shows Chaffinches arriving in south-east England and covers an area with a diameter of 300 kilometres (an area of more than 70,000 square kilometres). Often displays covering such wide areas give a very clear indication of particular lines of movement which can be investigated at larger scale – either by using the options for focussing in on smaller areas on the same set or by using another.

So far all the radar information being discussed is on PPI (Plan Position Indicator) displays. That is that a target detected may be at any height above the point on the ground where its blip indicates it is. These displays result from the familiar rotating aerials which sweep round and round receiving back signals from targets at any height. There are other sorts of aerial, particularly the ones that nod up and down, which are attached to sets which will give both an indication of where the target is in terms of range and direction but also its height above the ground. These are of particular interest in migration studies. Some of the early results obtained from south-east England showed that the median height of migration was about 600 metres on cloudy days and during cloudy nights. On clear days, that is with less than half the sky obscured by cloud, it was about half the height. On clear nights the median height for migrating birds was about 500 metres, only a little lower than in completely overcast conditions.

Radar technology has now developed to such an extent that it is not only possible to track individual birds but treatment of the signal received on reflection enables the 'signature' of the target to be traced on an oscilloscope. This changes as the shape of the bird's body alters with its wing beats (even

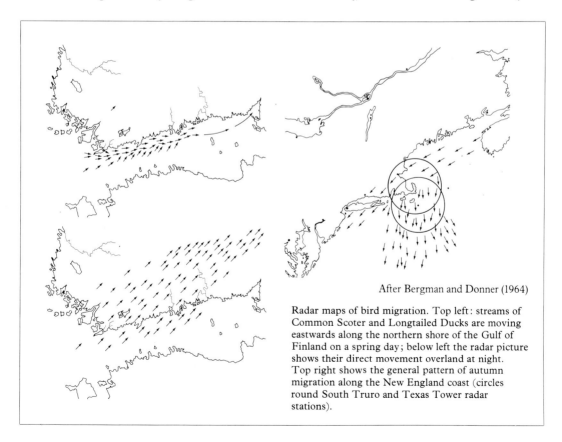

After Bergman and Donner (1964)

Radar maps of bird migration. Top left: streams of Common Scoter and Longtailed Ducks are moving eastwards along the northern shore of the Gulf of Finland on a spring day; below left the radar picture shows their direct movement overland at night. Top right shows the general pattern of autumn migration along the New England coast (circles round South Truro and Texas Tower radar stations).

its heart beats show on the most sensitive equipment). The changes seen are so great that it seems scarcely possible that simple muscular contraction could cause them but the answer lies in the 'target cross-section' mentioned earlier when comparing a head-on Starling with a jet aircraft. The bird's body reflects the radar pulses rather like a sphere of water but the strength of the reflected signal, for any particular wavelength of radar set, depends critically on the circumference of this sphere. This is particularly so if it lies in the region of one to ten wavelengths in size. The fluctuations are greatest between a circumference equal to the wavelength of the set and one 1.5 times the wavelength. The bigger sphere has an echoing area only a twentieth the smaller one. At twice the wavelength it is back to about two thirds of the maximum. These fluctuations decrease until steady surface scattering takes over at 30 to 40 wavelengths. These effects come into play because birds are just the right size to be in the areas of great fluctuation for many radar sets. The table (below) shows the cross-sections of some common European birds for 23 and 10 centimetre radars. Amazingly the strength of the echo which will be received back from some small birds is much larger than for some larger ones on 10 centimetre equipment.

An actual signature of the shape of the pulses received back from a flying bird that has been identified enables radar ornithologists to build up an identification library of the traces made on their sets by particular species. The sets are very different from the large long-distance surveillance radars used to track aircraft at considerable range. They are mostly developed from the radars used for anti-aircraft gun-laying. They operate over a short range and have target acquisition and locking systems so, once a 'target' has been tracked the aerial will automatically follow it. Used in daylight, with a telescope attached, this can enable birds seen at several hundreds of metre ranges to be identified. Other workers have simply released birds in front of the radar and then followed them. The photographs (see opposite) show the characteristic signatures received from waders, with their steady flapping flight, and a Redwing, with its bounding flight (the flat part of the trace is from the bird with its wings closed). This sort of radar observation has been used to follow natural migrants in many parts of the world and its pioneer, Glen Shaefer, has even used his portable set

Average masses and calculated radar cross-sections of some familiar birds for wavelengths of 23 and 10 cm.

Bird	Average mass (g)	Radius of equivalent sphere (r cm)	Relative radar sensitivity	
			23 cm	10 cm
Blackbird	100	2.88	42	8.7
Chaffinch	24	1.78	2.2	21
House Sparrow	25	1.81	2.4	21
Lapwing	200	3.63	91	58
Robin	17	1.60	1.1	18
Song thrush	70	2.56	22	2.1
Starling	80	2.67	29	34
Swift	43	2.17	8.3	13
warbler (large, garden warbler)	20	1.68	1.7	18
warbler (small, Chiff-chaff)	8	1.24	0.3	8.1

to investigate locust flights in north Africa. Its main problem is the relatively short range that birds may be followed – generally 3–5 kilometres each side of the set. However, this is more than countered by the precision with which the signature, height, speed and direction of the bird may be recorded. Very precise control of height has been observed along puzzling changes in direction. Also, happily for the birds, no damage is done to them by the focussed radio transmissions.

One further radar investigation needs to be mentioned. This involves a vertical set designed to quantify the amount of migrations taking place and the heights at which the birds are flying. This is very like the ceilometer technique already described, except that the radar can look through clouds and record the birds accurately. High resolu-

tion sets can be used since the maximum height at which small migrants are likely to be found is generally below 5 kilometres. As with all radar sets not specifically designed for very short range work there is a dead area. On most high resolution sets this area prevents birds below about 250 metres from being seen. This is dependent on the pulse-length which is generally of the order of a micro-second. Even amateur birders have been able to obtain suitable radar sets for this sort of investigation on the war surplus market and have used them to watch the nocturnal migration for themselves. It would, of course, help to know someone used to maintaining the same sort of set and any operator should take care to isolate his signal generation equipment from the neighbourhood television sets!

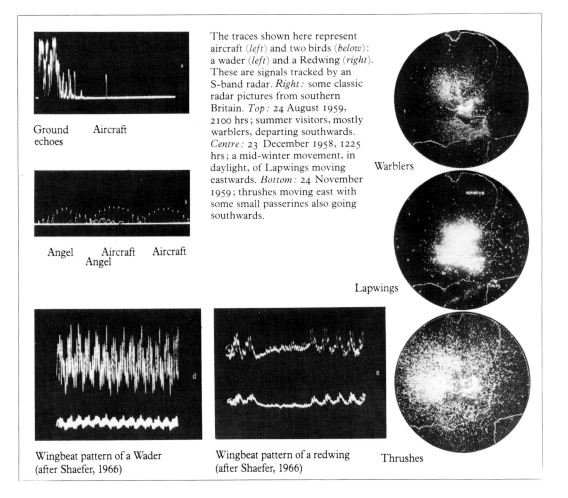

Ground echoes   Aircraft

Angel   Aircraft   Aircraft
      Angel

The traces shown here represent aircraft (*left*) and two birds (*below*): a wader (*left*) and a Redwing (*right*). These are signals tracked by an S-band radar. *Right:* some classic radar pictures from southern Britain. *Top:* 24 August 1959, 2100 hrs; summer visitors, mostly warblers, departing southwards. *Centre:* 23 December 1958, 1225 hrs; a mid-winter movement, in daylight, of Lapwings moving eastwards. *Bottom:* 24 November 1959; thrushes moving east with some small passerines also going southwards.

Warblers

Lapwings

Wingbeat pattern of a Wader (after Shaefer, 1966)

Wingbeat pattern of a redwing (after Shaefer, 1966)

Thrushes

Laboratory techniques

To many birders the sight of a bird in a cage is like the proverbial red rag to a bull. There are, however, many research workers who have made great advances on the mechanisms of migration from birds kept in cages under controlled conditions. Such investigators need to become very skilled aviculturalists for the birds to respond naturally to the stimuli which they receive. This applies equally to newly-taken wild birds which are tested in cages and then released as well as to birds raised in captivity and kept for very long periods. Almost anything is possible with caged birds. They may be taken from the nest in the wild and reared indoors in steel lined rooms so that there is no chance of them picking up visual clues from the sun, moon or stars nor experiencing the earth's natural magnetic forces. They may then be exposed to artificial skies in a planetarium or man-made magnetic fields within a coil and their reaction recorded and measured. The treatment of the bird forms the basis for the experiment but the recording of the bird's reaction provides the results. We have already seen, in chapter 5, some of the means of recording directional preferences of birds in cages, ranging from the very simple paper trace marked by ink on the bird's feet and plumage used by Stephen Emlen to the perches and micro-switches used by other workers. In one case it was proved that the birds only performed 'properly' if the perches were placed radially round the cage rather than tangentially on its circumference!

The manipulation of the bird in captivity, apart from the exclusion of known natural stimuli, may also include the upsetting of the natural day/night cycle to 'clock-shift' the birds. It may involve transporting them to an unfamiliar area and releasing them to see how they 'home' (this can be done for wild birds too but must not be lightly undertaken). It may involve interference with the bird's senses of sight, hearing or even taste or the upsetting of their normal hormonal balance either through the administration of drugs or surgery. Such drastic techniques are not now used very often and much more frequent today are experiments which involve the very detailed monitoring of the natural functions of the bird including weight, migratory activity, sexual activity and hormonal levels. As we have seen in earlier chapters it is becoming clear that many different facets of the bird's behaviour are involved in the control of migration.

Simulation

A valid study technique is to simulate what would happen if a theory were to be correct and then to test the results of the simulation against real life. We have already seen the results of Mengel's work on the species formation patterns in the New World wood warblers. It is, however, also possible to use a computer. You feed it with the known facts about likely weather, availability of food and winter goal as well as information on the optimal flight speed and daily time available for migratory flight, and ask, through careful programming, what would be the most efficient migration route. A few studies of this sort have been started but only one seems to have produced really useful results: Robert Eckhardt's simulation of the Arctic Tern movements from Labrador to the wintering area in the southern Atlantic and back again. He allowed the migrating term an average speed of 2.5 knots (roughly 6 kilometres an hour) day and night. He also fed into the computer information on the average wind speeds for the whole of the Atlantic. From this he calculated, allowing for the need to visit reasonable areas for feeding, the most efficient route that a tern might take. This was not based on the minimum distance but the minimum time. The maps (opposite) showing his results fit in very well with what is known of the actual routes that birds take from Labrador southwards.

Simulations which have taken into account prevailing weather conditions and topographical features have long been part of the research worker's tool-kit. Classic explanations of the concentrating effect of coastline and high-ground features on broadfront migration patterns readily explain the reasons for some of the world's best sited migration watching places. Arrivals of rare birds from distant areas may often be correlated with particular weather systems. Careful study of the wind strengths and directions within these systems can often show just how the passage of, for example North American land-birds found in Europe, may be assisted by a 'sling-shot' effect as they are whirled round active depressions. Indeed some of the most famous arrivals of American birds in Europe can be associated with particular storms which can be traced back to the spawning grounds of the major tropical storms off tropical South America. Similar arguments can be used to discover the likely paths of migrants going across the Atlantic from east to west and the arrival in western Europe of rare birds from central Asia.

A very unusual form of simulation has been pioneered by Bill Hale of Liverpool Polytechnic. He has taken winter specimens of Redshanks and measured their wing length,

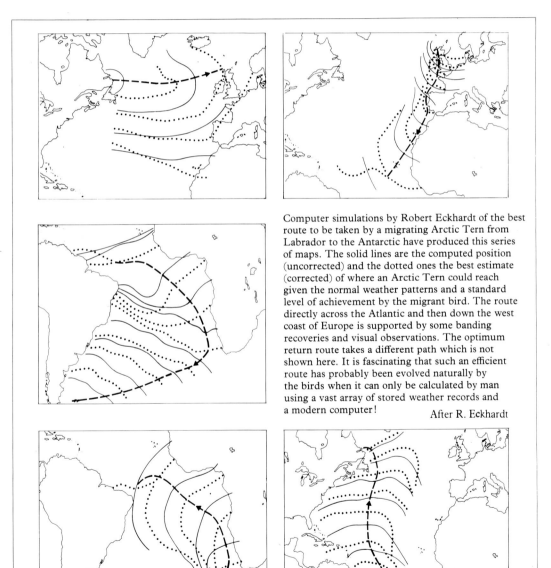

Computer simulations by Robert Eckhardt of the best route to be taken by a migrating Arctic Tern from Labrador to the Antarctic have produced this series of maps. The solid lines are the computed position (uncorrected) and the dotted ones the best estimate (corrected) of where an Arctic Tern could reach given the normal weather patterns and a standard level of achievement by the migrant bird. The route directly across the Atlantic and then down the west coast of Europe is supported by some banding recoveries and visual observations. The optimum return route takes a different path which is not shown here. It is fascinating that such an efficient route has probably been evolved naturally by the birds when it can only be calculated by man using a vast array of stored weather records and a modern computer!

After R. Eckhardt

tail length, bill length and tarsus length and width. He has also taken breeding season specimens and measured the same features. This data has been programmed into a computer which has then been asked the question: 'Which is the most likely breeding popula-tion from which each of these winter records originated?' Each answer provides him with a 'computer ringing recovery' which can then be used to plot the wintering area of the different breeding populations and even to plot the mean distance moved (taken as the

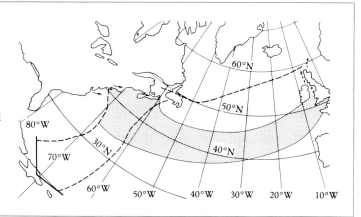

North American landbirds may be carried across the Atlantic by violent storms. Their 'normal' route southwards is shown by the arrow and the Atlantic storm track by the stipple. The dashed line between Newfoundland and Shetland is the northern limit of the storms which bring these small land migrants.

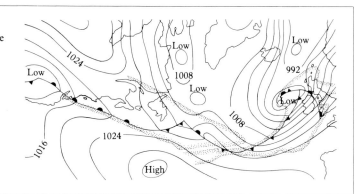

The weather situation at mid-day on 24th September 1975. The wave depression east of Newfoundland only took 30 hours to travel from Nova Scotia to the Irish coast. Four American landbirds were recorded in the next few days – Yellow-bellied Sapsucker, Black and White Warbler and Scarlet Tanager on Scilly Isles and a Scarlet Tanager in North Wales.

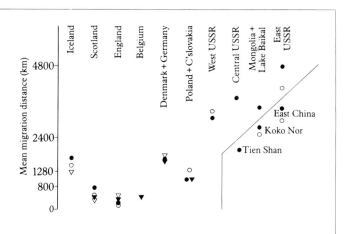

Using computer simulation, from detailed measurements of museum specimens, Bill Hale has constructed this chart of the distance moved on their winter migration by the different populations of Redshanks breeding from Iceland in the west (left) to the far east (right). Circles are computed data, triangles real ringing records: shaded symbols adults, open symbols first year birds.

After B. Hale (1973)

distance from the winter site to the centre of the breeding area).

Follow that bird!

Perhaps the most direct means of discovering migration patterns is for the research worker simply to follow the bird as it migrates. This simple idea has all sorts of practical problems. The birds migrate both day and night and so they may only be followed during the day unless they are marked in some way or illuminated by lights or radar. They tend to fly at low speeds relative to aircraft and take no account of flying restrictions round major airports and military installations. If they are chased by a helicopter or light aircraft at close range they tend to take avoiding action – so ruining the observations. Attempting to follow flying birds on land would normally be impossible, for their average flight speed would almost always be too high for a car to

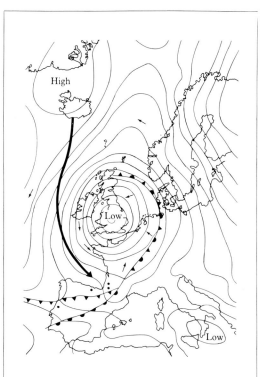

The strong southerly winds round the western edge of a depression centred over England assist migrant Redwings travelling from Iceland to their wintering grounds in south-west Europe. The four dots represent ringing recoveries. This weather map is a real one for mid-November 1959.

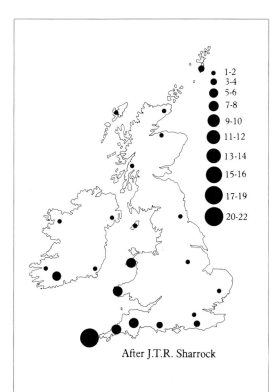

●	1-2
●	3-4
●	5-6
●	7-8
●	9-10
●	11-12
●	13-14
●	15-16
●	17-19
●	20-22

After J.T.R. Sharrock

Records of North American landbirds from Britain and Ireland during autumn and winter from 1958–72.

keep up. Following birds at sea is usually very difficult for they are likely either to be dynamic soaring species which may be attracted to the chasing ship or active flyers whose airspeed greatly exceeds anything but the most powerful ship.

All these problems are not enough to have discouraged all ornithologists from attempting to follow flying birds. There have been a few lucky observations from aircraft which have followed migrating flocks of large birds, like geese and swans, for short distances out of interest. However by far the best results have been from aircraft following radio-tagged birds: the bird does not have to be kept in sight all the time and the aircraft can be far enough from the bird not to disturb its normal behaviour. For example, a radio-tagged Tundra Swan was followed for over

11 hours in North America at an average ground speed of over 80 kilometres per hour. Similar techniques have been employed to track homing pigeons on release and were used, in a famous series of experiments where the tracking aerial was on a truck, to track Gray-cheeked Thrushes, Swainson's Thrushes and Veeries in Illinois. The range over which the transmitters could be heard was rather small and so the recording method was to track the bird from one position and then predict where it might be expected to cross a convenient road, drive quickly to that point and repeat the exercise.

The problems of actively following birds in the field can easily be dispensed with if the birds are very conspicuous and scarce. In such cases simultaneous observations over a wide area by observers, who record the birds accurately and at timed intervals, can produce successive observations of the same individual or flock. This can work particularly well with coasting seabirds where observers on successive headlands are quite likely to see identifiable flocks of birds moving along the coast.

Perhaps the classic case of successive observations concerns an artificially reared family of White Storks from Denmark which were traced into England and eventually to Cornwall during autumn 1971. All three were ringed with one recovered dead in Somerset and one eventually dying near Funchal in Madeira. This was a totally atypical journey for storks from Denmark to make (see map on page 209).

Statistics

The greatest ally of the research worker treating results that come from the research techniques detailed so far is statistics. It may be a daunting prospect for the naturalist in the field who enjoys bird watching but the usefulness of his observations, excluding general descriptions, rests on the skill of the analyst in using statistical techniques to find out the significance of the results obtained. Even the simple observations of daily numbers of birds in a census area need to be analysed to try and determine whether the observed increases and decreases might have

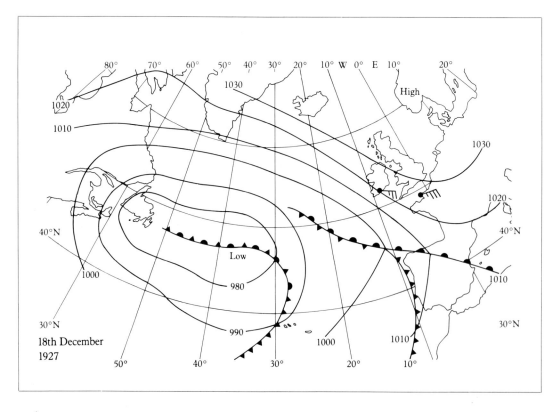

18th December 1927

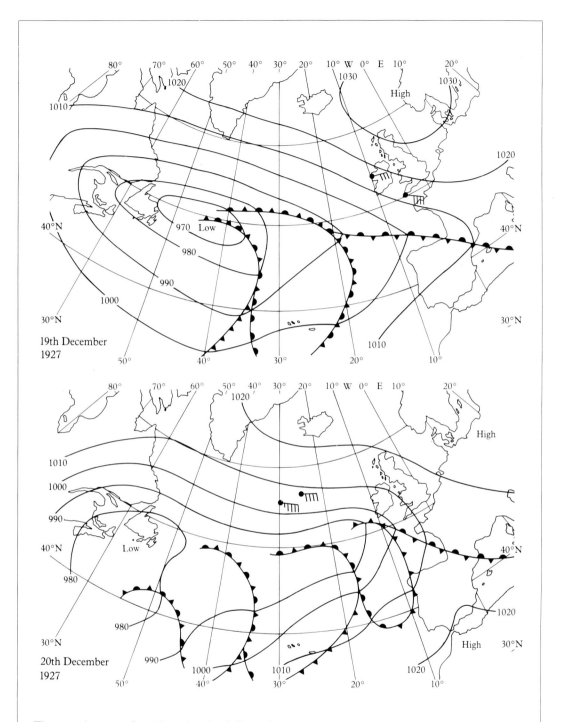

These weather maps, for 18th, 19th and 20th December 1927, show the freak storm conditions which allowed a ringed Lapwing from the Lake District (England) to reach Newfoundland. The bird had taken off to fly to warmer conditions in Ireland at the onset of cold weather. It obviously missed Ireland and, with tail winds of up to 45 knots along the top of the storm track, it was probably able to reach Newfoundland in well under two days flight time.

arisen from chance or do actually reflect real changes. Laboratory workers, where there are long series of observations of direction preferences in their birds, must know where the records do represent a real navigational ability or not. Analysts of ringing recoveries must be able to test whether apparent dif-ferences in recoveries from one year to another are real or just arise from chance.

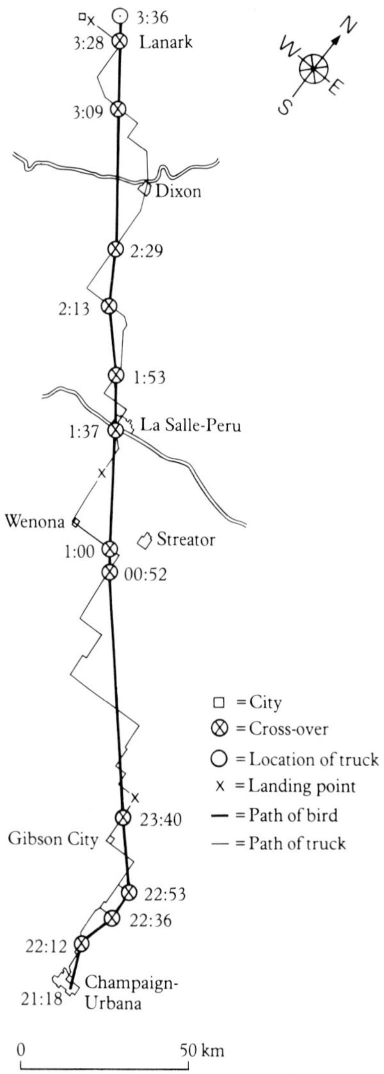

This is not the place to launch into a treatise on the methods used but it is a sobering thought that the analysis of migration data can be so complicated that it becomes the province of trained mathematicians who are also amateur birders! The problem of interpretation can be immense. For example Bird Observatory records in Britain are likely to be subject to a 'weekend' bias; the mainland stations are often better manned at weekends and so weekend records are more complete. An opposite bias exists on island observatories where the days when visitors change over, 'crossing days', are under-recorded! Weather conditions can be very different, even between two stations only 100 kilometres apart. A front which has just reached one observatory may produce a rush of grounded small migrants but the other, where the skies remained clear, may have no arrival at all. The aim of the analyst using the data may be to try to take out all these variations to compare gross numbers of birds moving from year to year – a daunting challenge but one being taken up by Keith Darby of the Mathematical Department at the University of Kent working on the British data.

The future

Over the next few decades much more will be learnt about the methods used for navigation and the physiological control of migration. Most of the new results which help our understanding of the underlying theories will come from detailed laboratory work on captive birds kept in artificial conditions. However, biologists working in the field, both amateur and professional, will continue to provide both basic descriptive information about migrants and also detailed records of numbers and dates which may allow the theories to be tested against 'real' birds. In the field, professional teams working with ever more miniaturized but powerful radio-tags will be gathering very detailed information on small numbers of individual birds. Amateurs and professionals co-operating to provide regular coverage of large areas will increasingly use colour rings, plumage dyes and wing-tags to

□ = City
⊗ = Cross-over
○ = Location of truck
x = Landing point
— = Path of bird
— = Path of truck

0   50 km

After Cochran, Montgomery and Graber (1967)

A radio-tagged Veery followed a very straight path for most of its flight but the tracking team was forced to zig-zag by the rectangular grid of roads!

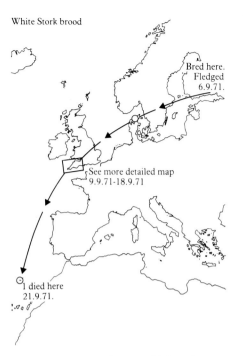

White Stork brood

Bred here.
Fledged
6.9.71.

See more detailed map
9.9.71-18.9.71

1 died here
21.9.71.

trace the exact routes used by specific populations of birds, particularly those which need conservation.

Ringers and banders throughout the region will continue marking migrants to get records of the birds in their winter quarters and on migration. Their work will continue the recent trend towards 'self-generated' data and, even within the various ringing schemes covering Europe, regular co-operation between the different countries will lead to many more controls of birds between countries. Whatever the state of professional research ordinary birdwatchers will continue to visit bird observatories and other observation points and count the moving birds. Others will simply marvel that the Swallow returning to its nest in the porch has flown to the tropics for the winter and returned to exactly the place from whence it started.

The large numbers of amateur birdwatchers can often provide information which, when it is combined, can tell a very interesting story. These two maps show the movements of a brood of storks ringed in Finland, which took a more westerly route than normal. The fate of these spectacular birds was meticulously recorded, particularly as they made their way across England.

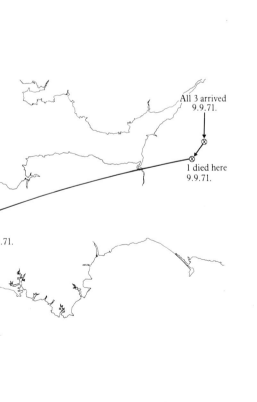

All 3 arrived
9.9.71.

1 died here
9.9.71.

Here 10-12.9.71.

11.9.71.

15-18.9.71.

12-13.9.71.

Observing migration for oneself

Migration is a very normal part of the lives of millions of birds throughout the world but is not necessarily visible to us, as human observers, in an obvious manner. Indeed for many people, living in cities in Europe or America, the most obvious bird movements will be the nightly roosting flights of masses of gulls and Starlings. The gulls move distances of 10 or 20 kilometres to roost on a river or reservoir whilst Starlings roost on tall buildings or in woods. The majority of the migrants will have moved at night unobserved and even those that do make their journeys during the day are often very much too high to be seen with ease. It is therefore very much easier to record migration by inference than by direct observation.

Keeping records

Any birdwatcher or country-lover makes a mental note during the spring of their first sighting of a swallow or warbler – species known to be summer migrants and absent for the long dark months of the winter.

Proper recording of migration is but an extension of this simple sort of observation. At a Bird Observatory there is generally a defined census area which is covered every day and has all the birds seen on it counted. The ordinary inland observer can easily choose to cover a convenient local area in just the same way. Many people will find this a fairly onerous task but, if it is simply being done for one's own satisfaction rather than as a chore for someone else it may quickly become a very enjoyable part of the day's routine. It is obviously crucial to choose a suitable area for the observations. It must be reasonably easy to reach and accessible in all weathers. If it is to be covered every day it must not be too large (it could be much larger if you are only going to do it every weekend) and, of course, it must contain suitable habitat to attract some migrant species.

For many years birdwatchers have regularly manned their own sites in this way. Commuters have walked the same, circuitous route, to the station each morning or have taken their lunch-break in the same parkland area. The dog has regularly been walked beside the local river or round a small reservoir so that his master could record the birds present or, sometimes over a period of several decades, the country parson has every morning taken the same walk round his local woods and fields and ended up with the morning reading of his personal weather station. Gilbert White, 200 years ago at Selborne in Hampshire, England, was really doing just that and hundreds like him have, over the years, provided the basis for local natural history observations of all sorts throughout much of Europe and America. In all such cases the results are much more satisfactory if the birds seen are counted, according to a properly thought out set of rules, and recorded systematically. The written record can then be kept over a period of years and consulted to point up differences between years and so on. However, this can become a chore and spoil the enjoyment of the actual watching which will, in any case, point up the gross ebb and flow of the bird populations.

Starting such a project from scratch does require a good deal of thought – once one has a couple of month's records accumulated it is very difficult to discard them in favour of a different area or recording method. So get it right first time! The first question of all is: are you going to try to do the site every day, most days or just at weekends? If the area is to be covered often on days when you are working then the time spent in reaching the site and covering it is likely to be really crucial, unless you are able to reorganize your working times. Morning visits, as soon after first light as possible, are generally best and so many people are unable to continue their recording right through the winter when they have to be at work before first light. In fact the great majority of migration takes place over a period of about three months in spring and three months in autumn – mostly March to May and August to October but this varies with latitude and country. These practical constraints may very severely limit the possible areas which can be covered during the week and it may be best to abandon the idea of daily recording but rather to choose a larger area to be done at weekends only.

The actual area chosen should have at least some habitats that are likely to be used by local populations of migrant species although, inevitably, it will only be a very exceptional

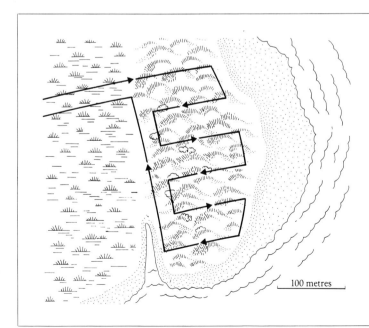

A small area of bushes on the shore might be surveyed each day by walking this route. Most birds should be seen and recorded by such a search and, provided it is standardised, this sort of daily census route – even over a small area inland – can produce a really useful series of daily records.

100 metres

site that might be regularly used by most of the migrant species. In general, areas of thick woodland should be avoided since it is so very difficult to count birds in this habitat. Open scrubland, pasture, grassland or damp areas particularly tend to attract a good variety of species but even formal gardens and open playing fields may have some migrants. Lakes, reservoirs and sewage works are particularly useful and many studies, inland, have been made of migrant waders attracted to such sites but hardly ever seen away from them. In many areas local groups of birdwatchers combine to provide as complete a coverage as possible through the different migration seasons. It is obviously quite easy to get this to work for flocks of ducks or groups of sandpipers at a reservoir or sewage farm but very much more difficult if the birds to be counted are small passerines in an area of bushes and trees. With the latter, the best chance of arriving at a comparable figure, day after day, is to have the same person doing the counting each time. Of course, if you are making your own record, doing the count in the same way each day is very important for this should diminish the variability of your results. It does not matter if all the birds present in the area are not being counted provided that the same proportion of each species is recorded each day. A standard walk through an area, recording only those birds encountered (a line transect) is just as acceptable, and probably easier to do, than a complete census. It is as well to plan the route to ensure that bright morning sunlight will not interfere with your observations and also so that crucial vistas over any areas of mud or water will not be obscured by the spring growth of leaves.

Having chosen the area what goes on record? The easy answer is everything. It is unlikely that the local House Sparrows are long distance migrants but their numbers may vary over the seasons and there will certainly be species, not normally thought of as migrants, which will move locally and fluctuate in numbers on your local patch. It is also a good idea to make one's own notes of the prevailing weather conditions rather than relying on gathering the weather data

later. The records can be written into an ordinary (large) diary but are much better recorded onto some sort of daily (or weekly) summary form with a column for each day's (or week's) observations and a line for each species. These can be purchased with the species names printed in as checklists for most areas but generally they have too few columns. Much better is the use of a loose-leaf system like that devised by the British Trust for Ornithology for use at Bird Observatories. Basically the observations are listed on a series of pages which are inserted within master-sheets which label each line with its species name. In addition there is also a daily sheet allowing extra observations to be noted (not of too much use for the private individual) and another system for recording any visible migration.

Some 20 years ago, before effective use of computers, a systematic nationwide system of gathering this sort of information was run by some enthusiasts of the B.T.O. Each day, with over 100 people sending in their records on an average of about 40 species, about 4000 records came in and the recording team was quickly swamped by data. The Inland Observation Posts scheme, as it was called, used a form with the species already printed in and was analysed by comparing the total for each day with the previous three to see whether a significant change in numbers had taken place. The scheme ground to a halt, choked by the data it was generating, but not before all sorts of changes in numbers of species previously thought of as sedentary or, at the most, short-distance migrants, had been demonstrated. A similar scheme launched now would rely on mark-sensing equipment to feed the results direct into a computer for processing rather than relying on a volunteer analyst!

The systematic records made at your chosen recording site will tell you much more than the usual notes kept by birdwatchers. Normally the first few records each year of migrants get recorded but it would be very difficult to try to work out the duration of passage from anything but systematically recorded data; the last dates of summer migrants in the autumn or winter migrants in the spring are often lost. Comparison between species may show how apparently unrelated birds move through your area at the same time – sometimes a coincidence but often a response to the same weather conditions. Even when systematic notes have been recorded the duration of passage is very often obscured by birds which remain to breed or winter.

It is also possible to observe migrating birds directly almost anywhere. At times they will be unmistakable – a flock of calling geese or wild swans flying over the centre of a large town or hordes of swallows beating low into a southerly wind in late September. The chances of seeing birds moving can vary greatly from place to place and at different times of the year, according to the weather conditions. First of all the sort of place to choose to look from should have a good unobstructed view – in a town an upstairs window may be good enough or the roof even better. In the country the brow of a hill overlooking lower ground is ideal. You should always try to look away from the sun or, at least, across it as it is very difficult to see birds if you are looking directly into the light. However, this is a distinct disadvantage in the spring when birds are moving from south to north; it obviously pays to pick them up before they reach you so that you can try to identify them when they are close to you. You should also remember that linear landscape features can act as concentrating lines

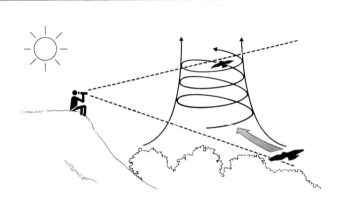

Watching birds in thermals (kettles) is best done from a rise since they are very often formed in country with tree cover. This can be a very rewarding form of bird-watching but the weather conditions must be right for the necessary lift to develop.

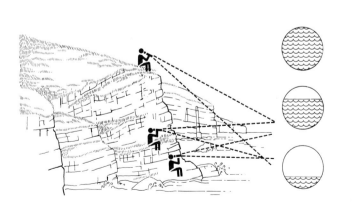

Watching birds soaring along a cliff or escarpment may best be done from slightly below the crest. This will depend on the height of the bird's flight and the direction that they are approaching the watching point.

for migrants, especially those that move during the day. In most cases they are ignored if they lie immediately across the migrant's path but are followed if they are diagonal to it. The most potent leading lines are ranges of hills, coastlines and valleys through high ground but some effect can be detected at the shore of large lakes or estuaries and even wide rivers. Many of the diurnal migrants call to each other as they fly and so a quiet place to watch from is a great advantage.

Weather conditions are of great importance. In the winter the very best and most spectacular day flights of all can be seen as cold weather invades from the north. Even in Central London thousands of birds like Lapwings, Redwings and Fieldfares as well as finches and buntings may be forced south by snow. There are many species only regularly recorded in the Central London area as they fly over under such conditions. At the main migration times the best flights are generally seen when the weather has been wet and windy for several days and then clears. If there is a contrary, but not very strong, wind then the migrants that have been held up will be moving in large numbers but low to escape the full adverse effect of the wind. Such conditions are particularly good for viewing goose, swallow, thrush and finch migration. In most instances, except when cold weather movements are involved, the morning from one hour to four hours after dawn will be the best time for watching. With very long-distance goose migration on regular routes, flocks may be due at particular places at predictable times of the day — different on spring and autumn migrations.

In fact the very best place to watch migration actually happening is undoubtedly on a coastal headland or small island. The land-bird migrants will often be very obvious since they will only be present on migration if there is no suitable breeding habitat. The great advantage is that the continuous passage of seabirds, often in huge numbers, may be visible from the shore. In fact, of course, the vast majority of migrating seabirds are well out to sea but many species can be seen from the shore and, under certain conditions, in really large numbers. Many species concentrate along exposed shorelines during storms and others, particularly terns and seaduck, may pass in huge numbers along south-facing shores in the spring and north-facing shores in the autumn. Where there is a choice the best sites for sea-watching may be found by trial and error – they will often differ for different weather conditions and will always be reasonably comfortable (the uncomfortable observer misses out on many of the passing birds). Recording the birds can be very tricky when hordes are moving and may belong to a dozen or more species. Many observers find that a sea-watching group needs a record keeper who hardly has time to look at the birds himself. Individuals, trying to do it by themselves, often resort to the use of a small tape-recorder and may have to concentrate on counting particular species rather than trying to keep notes on everything that is happening.

With moving birds (whether at coastal sites or inland) it is very important that the direction of movement is noted. This can often be done by arrows in the notebook (vertical means directly away if it points upwards and directly towards the observer if downwards). In many cases birds rounding a headland take a standard but changing course which can simply be indicated by their overall direction of movement. It has been generally agreed that the timing of movements may be very important and the unit used is generally the 'ten-minute block'. This means that the records for each ten minutes are kept separate and are related to an accurate watch. This enables the timing of movement to be compared between adjacent sites. It is always a good idea to time unusual observations, for migrating birds could always be seen somewhere else by another person who also times them. The story of the Storks in the previous chapter is a case in point.

All sorts of complications are introduced if observations are made from a moving boat. Not only are the birds on the move but so is the observer. In such cases the ten-minute block system is obviously necessary but so also is the recording of the ship's position. This can generally be obtained through the purser, even on a busy ferry, for two or three points each day and simple calculations can then be performed to position all the observations. It is always a very good idea to try to separate out three different categories of bird seen from moving boats. There are those 'associated' with the boat, generally gulls in northern waters, hanging round the stern waiting for any scraps. Secondly, there are the static birds, usually sitting on the water that the boat passes by. Lastly, there are the actively 'moving' birds on migration whose direction of movement should really be recorded too. Most ocean areas have special surveys being conducted which are attempting to summarize the distribution of birds at sea. In areas like the North Sea this is particularly important as oil exploration and exploitation inevitably increases the risk of pollution.

A very special sort of observation can be made from static points at sea. In the past these used to be the Ocean Weather Ships which kept station, often vast distances from even the main shipping routes, but modern satellite imaging and recording has meant that the records that they used to provide can now be gathered remotely. Lightships are also being superseded by automatic stations but oil installations are becoming more and more common with quite large crews, often including birders. Indeed the North Sea has its own Bird Club with over 100 members who are actively birdwatching

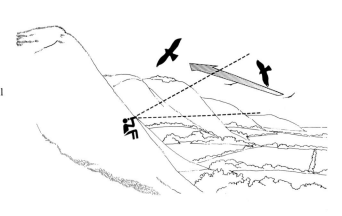

Sea-watching sites are always a compromise between height, for range of vision, and lowness, to increase the field seen in the binoculars. The middle position is best: too low and birds will pass below the horizon: too high and only a little of the passage will be visible at any one time.

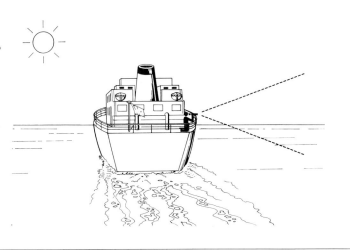

Birding from a boat should never be into the sun and should take place from a reasonable height in a sheltered position, not subject to too much vibration or rolling.

offshore. To the land-based observer it sounds an easy life but to anyone who has been offshore the difficulties they encounter will readily be appreciated. Small land-birds tend to huddle into nooks and crannies in the complicated pipe-work of the installation allowing only brief glimpses. Their behaviour is severely modified by their surroundings and so plays little part in identification and, in any case, the vast range of species that might occur makes accurate recording a very difficult problem. Many rare migrants have been recorded and they are sometimes so exhausted they allow themselves to be picked up. They are then sent ashore on the next helicopter and astonished birders, ashore,

have been presented with birds as rare as the Lanceolated Warbler in a small cardboard box from the kitchen of a rig! On most installations small numbers of migrants are present throughout the migration season. In foggy weather hundreds, even thousands, may appear; flares or floodlights are also an attraction to nocturnal migrants. Most of these will only stay for a few hours but there can be a considerable mortality if the conditions remain bad for several days. Such dying birds would probably have perished even sooner if there was not the rig for them to rest on.

Equipment

Birdwatching does not necessarily need any equipment but, in almost all circumstances, a pair of binoculars is a great help. All sorts of advice can be given about their purchase but the most important thing is that they should be comfortable to use and *available*: the best advice to the beginner might even be to burn his binocular case as soon as he purchases the binoculars. More birds are missed or not identified whilst the would-be watcher struggles to extract his binoculars from their case than for any other reason. Generally 8 × to 10 × magnification should be chosen. If the magnification is less, the glasses are not powerful enough and if more powerful it is difficult to hold them steady. No-one, except for specialist reasons, should contemplate hand-holding any binoculars with a magnification of more than 12 ×. With the conventional design an 8 × 30, 9 × 40 or 10 × 50 pair should give good results provided that it has been well-made. Always go to a reputable dealer, look through the different makes and be advised (to a certain extent) by what the salesman has to say. In Britain, adequate and cheap binoculars are available from Japan and Russia. Excellent, traditional designs are also available from East Germany. Other makes, at up to four times as much

Opposite A party of bird watchers observing a flock of White Pelicans in Turkey. The migration route followed by many birds along the east coast of the Mediterranean is an excellent place to observe migration and attracts large numbers of both bird watchers and birds.

provide surprisingly little improvement optically. There has been, however, one great improvement within the last decade – the introduction of the roof-prism design. These 10 × binoculars with really bright images can be as light (or lighter) than the conventional 8 × 30 but the cost is great for they are precision-made instruments.

In the past all serious birdwatchers used to travel around with massive, brass-bound three or four-draw telescopes. After quite a time they are coming back into general use – at least amongst the most ardent birders. Two sorts are used. In Britain the prismatic spotting scope is most often used with a variable magnification of 15 × to 60 ×. In America the fat, squat shape of the mirror scope is becoming more and more common. These instruments provide magnifications absolutely unobtainable using the conventional prismatic design but require a very firm tripod and a great deal of practice in their use.

Many birders are also to be found festooned with cameras and tape-recorders but these are not necessary aids for migration watching. What is needed is a good bird-book – or more properly a selection of good books. If you are in your home area and have been birding there for many years you may not need to consult any fieldguide from one year to the next but the chances are that you will constantly be referring to them to see whether the exact plumage on a bird you have seen is illustrated or not. Most parts of the world have a guide relevant to the area and often a combination of several is needed. The bibliography (see page 220) lists the most useful guides and the areas covered.

However, the guides can only be used if you see the birds properly and are able to note what you see in such a way that it relates to the illustration and descriptions in the guide. Techniques of birdwatching vary from habitat to habitat and even place to place but there are a number of principles that stand one in good stead everywhere. Unlike watching mammals, where the wind direction is of such great importance as it may carry your scent, the most important point with birds is to try not to look at them against the light.

If you have to then you will see why: all detail and colour tends to disappear leaving a silhouette. In fact the most common cause for alarm amongst birds is a sudden movement or flash of reflection from a shining lens. If you wish to get close to static birds take it easy and keep off the skyline; it will help if you advance on them from behind a tree or other obstruction. However, in many instances the best bet is simply to wait, in a still and comfortable position, for the birds to come to you. This will work in woodland or scrubland areas where it is often easy to work out which way a flock of birds is working and wait for them. It is also very easy if there is a food source or, more likely, a drinking place that the birds are visiting. Under these circumstances small birds may come within a few feet of a sitting birdwatcher who keeps relatively still but is otherwise not concealed.

For the really lazy birdwatcher a car can provide the ideal hide or blind. In many areas an early morning drive along the coast will reveal what has arrived during the night on the tidal pools and along the shore-line with little effort. In America, but not in Europe, a wide variety of species respond to 'pishing': the making of high pitched squeaking and hissing noises through pursed lips or by noisily kissing the back of one's hand. This is particularly satisfactory when a mixed group of wood-warblers flitting around in thick cover 30 feet above one's head can be persuaded to ground level where, on fairly open perches, they can be identified using a field guide. In Britain I have had mixed flocks of tits respond in the autumn and also a few individual warblers but the only consistent interest is shown by nesting adult tits at about the time their offspring have left the nest.

Any identification you put on a bird remains between you and your conscience *until* you start to submit records to other people, whether a formal report is involved or you are simply cooperating in watching an area. Once you start to do this you should always be prepared to defend any identification which you claim by the provision of a description of the bird you saw and not what you have since found out from the book. There are,

you will find, systems of records committees which consider such reports and it is up to you to provide descriptions as required if you wish your records to be accepted.

More advanced techniques

A particular breed of migration watcher gets hooked on moonwatching after seeing the first bird apparently swimming over the bright disc which his watering eyes have been staring at for the previous 20 minutes. Others breathe a sigh of relief that they can actually say that the technique works and never try it again! Bearing this in mind I would recommend everyone to have a go. It is best done by moving a camp-bed into the garden on warm autumn evenings or even opening a bedroom window and doing it from indoors. A tripod will be needed, whether binoculars or a telescope is used, and the moon needs to be at least $\frac{3}{4}$ full and high in the sky. If you wish to calculate the actual track of the birds or the numbers passing consult the moonwatching tables.

On overcast nights it is almost as interesting to go outside and listen to the sounds made by the migrants. Several species make quite noisy contact calls as they fly through the night, for example geese, swans and many of the thrush species. In particular when there is foggy weather many species of wader call a great deal and it is often possible, even inland in England, to hear six or eight species on a single August or September night. For this and for moonwatching, autumn is much better than spring for there are many more birds moving after the breeding season than before it. It also seems likely that the young birds call more frequently than their parents.

Of course much of the information on orientation and physiology must come from professional ornithologists working full-time. However, there is still one field where the amateur has a real role to play and that is in bird-ringing. If you wish to take this up you will have to undergo training with an experienced ringer before being able to ring by yourself. In almost all countries the catching and ringing of birds is also controlled by legislation and so various permits and licences will be needed. Ringers (called banders in

America) always hope for the long-distance recoveries of their birds but these are very few and far between. A few days ago I learnt of a Chiffchaff I had ringed in Hertfordshire which had been found in Mali, Africa. This is the first recovery I have personally had of a warbler or flycatcher south of the Sahara from some 25,000 or more ringed over a period of 21 years!

However, it is quite possible to study migration without needing to have recoveries of the birds at a distance. Ringing enables the same individual to be identified throughout its life. Thus the timing of the return of migrants can be recorded from year to year, their weight increases and losses logged and the relation between their departure time and moult period studied. In the long-run the most important point about migration is that it is a survival strategy for the species as a whole and ringing studies enable the bird's survival and mortality patterns to be plotted.

Where to go

Many birdwatchers like to spend their holidays watching birds and studying migration. They can do this in many different ways, ranging from staying at hotels on the coast, spending their days watching birds, to mounting an expedition to spend a long period camping on a remote and uninhabited island. A happy compromise, interesting both for the beginner and the expert, is to visit a Bird Observatory. Many of these are on islands and headlands round the coast but some are by lakes, like Long Point on the Great Lakes, and others are on the high ground, like Col de Bretolet, an Alpine pass. Most observatories have a permanent staff who are able to provide advice and, if it is needed, tuition for the visitors who are encouraged to participate in the routine work of the station.

The standard of accommodation varies greatly from place to place. At the most luxurious, meals will be provided and the overall impression will be of a smoothly run guest house or small hotel. At others those staying will be expected to bring all their own food and cook it and, since the crossing is unpredictable, to have some additional rations with them too. The observatory, in these cases, generally provides a roof over one's head, a bed and some cooking facilities.

If you wish to see as many migrants as possible the best chance is to join in with the local 'listers' or 'twitchers'. They will have a formal (Rare Bird Alert) system of notification of the location of rare birds or an informal 'grapevine' which disseminates the same information more casually. Another course is to join the local birdwatching society and to go out on field meetings which will be led by an experienced birder who will be able to help you with identification and fieldcraft problems. Whatever you do you should always remember that the welfare of the bird should be the main priority at all times. There have been horrifying stories of rare migrants being hounded from bush to bush, even roused from their roost during the night. Tired migrants are easily observed from a reasonable distance and should not be crowded so that they are unable to feed or rest. As a second priority the birdwatcher should always respect private property and not trample crops or damage fences or hedges to try and see a rare bird.

Many people will wish to go abroad to look at migration. The listed observatories will provide an excellent holiday whether far from home with all the birds new to the observer or near at hand with many of them familiar species. There are also many tours arranged by specialist travel agencies with experienced leaders. These may go to such sites as the Bosphorus, Gibraltar or even the Caspian Sea. They might also simply be normal tours at migration time to areas where large numbers of birds pass through. Many people will find such tours very great fun. These tours, however, should always be approached seriously for it would be most unfair on the people participating who were really keen if uninterested people also book on it. I have met very few birdwatchers who have gone on such tours who have not thoroughly enjoyed them and many use them year after year.

Bibliography

Baker, Robin *The Evolutionary Ecology of Animal Migration* Hodder & Stoughton, 1978.

Baker, Robin *The Mystery of Migration* Macdonald, 1980.

Bernis, Francisco *Aves Migradoras Ibericas* SEO, Madrid, 1966 (4 volumes), 1967 and 1970.

Bernis, Francisco *Migracion en Aves* SEO, Madrid, 1966.

Cramp, Stanley *et al.* (Eds). *Handbook of the Birds of Europe, the Middle East and North Africa* Oxford University Press, 1977, 1980 & 1983 (continuing).

Dorst, J. *The Migration of Birds* Heinemann, 1962.

Durman, Roger *Bird Observatories in Britain and Ireland* Poyser, Calton, 1976.

Eastwood, E. *Radar Ornithology* Methuen, 1967.

Edelstam, C. *The Visible Migration of Birds at Ottenby: Sweden* Vår Fågelvärld: Supp. 7: 1972.

Farner, D. S. and King, J. R. (Eds). *Avian Biology:* volume 5 Academic Press, 1975.

Glutz von Blotzheim, U. N. *et al.* (Eds). *Handbuch der Vögel Mitteleuropas* Akademische Verlagsgesellschaft, 1966 – continuing.

Keast, Allan & Morton, Eugene S. (Eds) *Migrant Birds in the Neotropics.* Smithsonian Institution Press, 1980.

Lack, David *Ecological Isolation in Birds* Oxford University Press, 1971.

Matthews, G. V. T. *Bird Migration* (2nd edition) Cambridge University Press, 1968.

McClure, H. E. *Migration and Survival of the Birds of Asia* U.S.A.M.C., Bangkok, 1974.

Mead, Chris *Bird Ringing,* BTO, Tring, 1974.

Moreau, R. E. *The Palaearctic-African bird migration systems.* Academic Press, 1972.

Palmer, R. S. (Ed). *Handbook of North American Birds.* Yale University Press, 1962, 1976 (continuing).

Pennycuick, Colin *Animal Flight.* Edward Arnold, 1972.

Riddiford, Nick and Findley, Peter *Seasonal Movements of Summer Migrants.* BTO, Tring, 1981.

Schmidt-Koenig, K. and Keeton, W. T. (Eds). *Animal Migration, Navigation and Homing.* Springer-Verlag, 1978.

Schmidt-Koenig, Klaus *Avian Orientation and Navigation.* Academic Press, 1979.

Schüz, E. *Grundries der Vogelzugskunde.* Paul Parey, Berlin, 1971.

Ulfstrand, S. *et al. Visible Bird Migration at Falsterbo, Sweden.* Vår Fågelvärld: Supp. 8, 1974.

Zink, G. *Der Zug europäischer Singvögel* (An Atlas of ringed birds). Vogelzug-Verlag: 1973, 1975, 1982 (3 volumes so far published).

Journals
American Birds National Audubon Society. U.S.A.

Animal Behaviour Association for the Study of Animal Behaviour. International.

Ardea Dutch Ornithologists Union.

Ardeola Spanish Ornithological Society.

Auk American Ornithologists Union.

Bird Study and also *Ringing and Migration* British Trust for Ornithology.

British Birds Independent magazine. Britain.

Condor Cooper Ornithological Society, U.S.A.

Ibis British Ornithologists Union.

Journal of Field Ornithology (formerly *Bird Banding*) N.E. Bird Banding Association, U.S.A.

Journal für Ornithologie German Ornithological Society.

Vår Fågelvärld Swedish Ornithological Society.

Wilson Bulletin Wilson Ornithological Society. U.S.A.

List of scientific names

Greater Northern Diver (Common Loon) *Gavia immer*
Horned (Slavonian) Grebe *Podiceps auritus*
Great-crested Grebe *P. cristatus*
Red-necked Grebe *P. grisegena*
Western Grebe *Aechmophorus occidentalis*
Short-tailed Albatross *Diomedea albatrus*
Laysan Albatross *D. immutabilis*
Black-footed Albatross *D. nigripes*
Northern Fulmar *Fulmarus glacialis*
Greater Shearwater *Puffinus gravis*
Sooty Shearwater *P. griseus*
Manx Shearwater *P. puffinus*
Balearic Shearwater *P. p. mauretanicus*

Storm-Petrel *Hydrobatus pelagicus*
Leach's Storm-Petrel *Oceanodroma leucorhoa*
Least Storm-Petrel *Halocyptena microsoma*
European White Pelican *Pelecanus onocrotalus*
Gannet *Sula bassana*
Great Cormorant *Phalacrocorax carbo*
Cattle Egret *Bubulcus ibis*
American Bittern *Botaurus lentiginosus*
White Stork *Ciconia ciconia*
Lesser Flamingo *Phoeniconaias minor*
Whistling (Bewick's) Swan *Cygnus columbianus*
Whooper Swan *C. cygnus*
Mute Swan *C. olor*
White-fronted Goose *Anser albifrons*
Graylag Goose *A. anser*
Pink-footed Goose *A. brachyrhynchus*
Snow (Blue) Goose *A. caerulescens*
Ross's Goose *A. rossii*
Brent (Brant) Goose *Branta bernicla*
Canada Goose *B. canadensis*
Barnacle Goose *B. leucopsis*
Red-breasted Goose *B. ruficollis*
Shelduck *Tadorna tadorna*
Pintail *Anas acuta*
American Wigeon *A. americana*
Cape Wigeon *A. capensis*
Northern Shoveler *A. clypeata*
Blue-winged Teal *A. discors*
Eurasian Wigeon *A. penelope*
Mallard *A. platyrhynchos*
Garganey *A. querquedula*
Ring-necked Duck *Aythya collaris*
Common Pochard *A. ferina*
Tufted Duck *A. fuligula*
Canvasback *A. valisineria*
Black Scoter *Melanitta nigra*
Oldsquaw *Clangula hyemalis*
California Condor *Vultur californianus*
Honey Buzzard *Pernis apivorus*
Snail Kite *Rostrhamus sociabilis*

Black Kite *Milvus migrans*
Red Kite *M. milvus*
Indian White-backed Vulture *Gyps bengalensis*
Short-toed Eagle *Circaetus gallicus*
Marsh Hawk (Hen Harrier) *Circus cyaneus*
Sharp-shinned Hawk *Accipiter striatus*
Common Buzzard *Buteo buteo*
Rough-legged Hawk *B. lagopus*
Broad-winged Hawk *B. platypterus*
Swainson's Hawk *B. swainsoni*
Osprey *Pandion haliaetus*
Merlin *Falco columbarius*
Eleonora's Falcon *F. eleonorae*
Peregrine Falcon *F. peregrinus*
Common Kestrel *F. tinnunculus*
Red-footed Falcon *F. vespertinus*
Rock Ptarmigan *Lagopus mutus*
Gray Partridge *Perdix perdix*
Common Quail *Coturnix coturnix*
Whooping Crane *Grus americana*
Sandhill Crane *G. canadensis*
Crane *G. grus*
Japanese Crane *G. japonensis*
Siberian White Crane *G. leucogeranus*
Water Rail *Rallus aquaticus*
Corncrake *Crex crex*
Sora *Porzana carolina*
Lapwing *Vanellus vanellus*

Eurasian Golden Plover *Pluvialis apricaria*
American Golden Plover *P. dominica*
Ringed Plover *Charadrius hiaticula*
Semipalmated Plover *C. semipalmatus*
Whimbrel *Numenius phaeopus*
Bristle-thighed Curlew *N. tahitiensis*
Redshank *Tringa totanus*
Common Snipe *Gallinago gallinago.*
Eurasian Woodcock *Scolopax rusticola*
Sanderling *Calidris alba*

Red Knot *C. canutus*
Pectoral Sandpiper *C. melanotos*
Ruff *Philomachus pugnax*
Long-tailed Skua (Long-tailed Jaeger) *Stercorarius longicaudus*
Parasitic Jaeger *S. parasiticus*
Pomarine Skua (Pomarine Jaeger) *S. pomarinus*
Herring Gull *Larus argentatus*
Red-legged Kittiwake *L. brevirostris*
Lesser Black-backed Gull *L. fuscus*
Bonaparte's Gull *L. philadelphia*
Sabine's Gull *L. sabini*
Black-legged Kittiwake *L. tridactylus*
Aleutian Tern *Sterna aleutica*
Common Tern *S. hirundo*
Arctic Tern *S. paradisaea*
Little Auk (Dovekie) *Plotus alle*
Razorbill *Alca torda*
Guillemot (Common Murre) *Uria aalge*
Tufted Puffin *Lunda cirrhata*
Rock Dove *Columbia livia*
Wood Pigeon *C. palumbus*
Collared Turtle Dove *Streptopelia decaocto*
Turtle Dove *S. turtur*
Cuckoo *Cuculus canorus*

Yellow-billed Cuckoo *Coccyzus americanus*
Black-billed Cuckoo *C. erythropthalmus*
Roadrunner *Geococcyx californianus*
Scops-Owl *Otus scops*
Short-eared Owl *Asio flammeus*
Long-eared Owl *A. otus*
Lesser Nighthawk *Chordeiles acutipennis*
Poorwill *Phalaenoptilus nuttallii*
Eurasian Nightjar *Caprimulgus europaeus*
Chimney Swift *Chaetura pelagica*
Common Swift *Apus apus*
Pallid Swift *A. pallidus*

Ruby-throated Hummingbird *Archilochus colubris*
Hoopoe *Upupa epops*
Wryneck *Jynx torquilla*
Common Flicker *Colaptes auratus*
Yellow-bellied Sapsucker *Sphyrapicus varius*
Acadian Flycatcher *Empidonax virescens*
Skylark *Alauda arvensis*
Horned Lark (Shorelark) *Eremophila alpestris*
Rough-winged Swallow *Stelgidopteryx ruficollis*
Sand Martin (Bank Swallow) *Riparia riparia*
Crag Martin *Hirundo rupestris*
Swallow (Barn Swallow) *H. rustica*
Cliff Swallow *Petrochelidon pyrrhonota*
House Martin *Delichon urbica*
White Wagtail *Motacilla alba*
Pied Wagtail *M. a. yarrellii*
Gray Wagtail *M. cinerea*
Yellow Wagtail *M. flava*
Red-throated Pipit *Anthus cervinus*
Tree Pipit *A. trivialis*
Red-backed Shrike *Lanius collurio*
Northern Shrike *L. excubitor*
Woodchat-Shrike *L. senator*
Bohemian Waxwing *Bombycilla garrulus*
Dipper *Cinclus cinclus*
Carolina Wren *Thyrothorus ludovicianus*
Mockingbird *Mimus polyglottos*
Alpine Accentor *Prunella collaris*
Dunnock *P. modularis*
Eurasian Robin *Erithacus rubecula*
Bluethroat *Luscinia svecica*
Redstart *Phoenicurus phoenicurus*
Whinchat *Saxicola rubetra*
Stonechat *S. torquata*
Wheatear *Oenanthe oenanthe*
Veery *Catharus fuscescens*
Hermit Thrush *C. guttatus*

Gray-cheeked Thrush *C. minimus*
Swainson's Thrush *C. ustulatus*
Redwing *Turdus iliacus*
Blackbird *T. merula*
American Robin *T. migratorius*
Song Thrush *T. philomelos*
Fieldfare *T. pilaris*
Bearded Reedling *Panurus biarmicus*
Cetti's Warbler *Cettia cetti*
Lanceolated Warbler *Locustella lanceolata*
Savi's Warbler *L. luscinioides*
Grasshopper Warbler *L. naevia*
Paddy-Field Warbler *Acrocephalus agricola*
Moustached Warbler *A. melanopogon*
Marsh Warbler *A. palustris*
Sedge Warbler *A. schoenobaenus*
Reed Warbler *A. scirpaceus*
Icterine Warbler *Hippolais icterina*
Melodious Warbler *H. polyglotta*
Blackcap *Sylvia atricapilla*
Garden Warbler *S. borin*
Subalpine Warbler *S. cantillans*
Whitethroat *S. communis*
Lesser Whitethroat *S. curruca*
Menetries's Warbler *S. mystacea*
Dartford Warbler *S. undata*
Bonelli's Warbler *Phylloscopus bonelli*
Arctic Warbler *P. borealis*
Chiffchaff *P. collybita*
Pallas's Leaf Warbler *P. proregulus*
Wood Warbler *P. sibilatrix*
Willow Warbler *P. trochilus*
Ruby-crowned Kinglet *Regulus calendula*
Firecrest *R. ignicapillus*
Goldcrest *R. regulus*
Golden-crowned Kinglet *R. satrapa*
Brown Creeper *Finschia novaeseelandiae*
Collared Flycatcher *Ficedula albicollis*
Pied Flycatcher *F. hypoleuca*
Spotted Flycatcher *Muscicapa striata*
Blue Tit *Parus caeruleus*
Great Tit *P. major*
Common Tree-Creeper *Certhia familiaris*
Ortolan Bunting *Emberiza hortulana*
Snow Bunting *Plectrophenax nivalis*
Fox Sparrow *Zonotrichia iliaca (*replaces *Passerella)*
Savannah Sparrow *Ammodramus sandwichensis*
Indigo Bunting *Passerina cyanea*
Western Tanager *Piranga ludoviciana*
Scarlet Tanager *P. olivacea*
Black-and-White Warbler *Mniotilta varia*
Orange-crowned Warbler *Vermivora celata*
Nashville Warbler *V. ruficapilla*
Yellow-rumped Warbler *Dendroica coronata*
Blackburnian Warbler *D. fusca*
Kirtland's Warbler *D. kirtlandii*
Chestnut-sided Warbler *D. pensylvanica*
Yellow Warbler *D. petechia*
Blackpoll Warbler *D. striata*
Black-throated Green Warbler *D. virens*
American Redstart *Setophaga ruticilla*
Northern Waterthrush *Seiurus noveboracensis*
Mourning Warbler *Geothlypis philadelphia*
Wilson's Warbler *Wilsonia pusilla*

White-eyed Vireo *Vireo griseus*
Hutton's Vireo *V. huttoni*
Red-eyed Vireo *V. olivaceus*
Orchard Oreole *Icterus spurius*
Western Meadowlark *Sturnella neglecta*
Rusty Blackbird *Euphagus carolinus*
Bobolink *Dolichonyx oryzivorus*
Chaffinch *Fringilla coelebs*
Brambling *F. montifringilla*
Serin *Serinus serinus*
Eurasian Goldfinch *Carduelis carduelis*

Eurasian Siskin *C. spinus*
American Goldfinch *C. tristis*
Linnet *Acanthis cannabina*
Red Crossbill *Loxia curvirostra*
Spanish Sparrow *Passer hispaniolensis*
Common Starling *Sturnus vulgaris*
Golden Oriole *Oriolus oriolus*
Raven *Corvus corax*

221

Index